Multiple-choice Questions in Basic Sciences for the MRCPsych Part II

Multiple-choice Questions in Basic Sciences for the MRCPsych Part II

Sudip Sikdar, MBBS, MD, MRCPsych

Specialist Registrar in Psychiatry
Fazakerley Hospital, Longmoor Lane, Liverpool L9 7AL, UK

 PETROC PRESS

Petroc Press, an imprint of LibraPharm Limited

Distributors

Plymbridge Distributors Limited, Plymbridge House, Estover Road, Plymouth PL6 7PZ, UK

First edition 1999
Reprinted 1999

Published in the United Kingdom by
LibraPharm Limited
3b Thames Court
High Street
Goring-on-Thames
READING
Berks
RG8 9AQ
UK

A catalogue record for this book is available from the British Library

ISBN 1 900603 56 X

Printed and bound in the United Kingdom by
MPG Books Limited, Bodmin, Cornwall PL31 1EG

Contents

Preface

The idea of this book was forming when I was writing the Part II examination. I could not find a single comprehensive book that covered solely the basic sciences part of the examination, and hence I struggled greatly preparing it. Though there are standard recommended textbooks for the clinical paper, there is no such common ground for the basic sciences paper. To make matters worse, the examination does not, as yet, fully reflect the curriculum, some topics being examined in detail while others are covered only superficially if at all. Rather than mixing up different questions from different chapters of the syllabus, I have deliberately kept them separate, hoping that it will help in the revision of particular chapters. The number of questions in each chapter is intended to reflect the importance of the topic in the examination. I have also deliberately kept the explanatory answers brief, with the intention of encouraging the student to read in greater depth around the topics in question. Because of the general vagueness of the basic sciences paper, it is extremely difficult to reflect exactly the real examination questions, and I must admit that this book does not attempt to do the impossible. However, I hope that it will help in preparation for this difficult part of the examination.

Liverpool, 1998   S.S.

1. Human Development

Q1.1 A child of 3 years:

A. Can ride a tricycle
B. Can give his/her full name and sex
C. Can copy a cross
D. Can put on shoes
E. Can name at least one colour

Q1.2 The following statements are true about attachment behaviour:

A. Attachment of mothers to infants occurs crucially during the early hours following birth
B. Studies have failed to find an increase in the strength of mothering behaviour as a result of extended contact with the infant
C. Stranger anxiety is a necessary component of attachment behaviour
D. Onset of attachment behaviour coincides with development of object permanence
E. Avoidant attachment in childhood is often associated with antisocial behaviour in adult life

Q1.3 The following statements are true about neonatal behaviour:

A. Rudimentary stepping behaviour in neonates disappears by 2 months
B. Truly intentional social behaviour appears around 6 months
C. Truly reciprocal social behaviour appears around 6 months
D. *Motherese* refers to modification of a mother's facial expression when interacting with her child
E. Adult level of vision is reached by 8 months

Q1.4 The following statements are true of Thomas and Chess's NYLS (New York Longitudinal Study) of temperament:

A. It was based on the assumption that temperament is largely genetically determined
B. It used the Child Behaviour Checklist to formulate the dimensions of temperament
C. It recruited the sample from a homogeneous middle-class group of parents
D. Activity levels accounted for the greatest variance between children
E. Slow to warm up babies were predisposed to developing later behaviour disorder

Q1.5   'Resilience in the face of risk' was attributed by Thomas and Chess to:

A. Development of schizoid personality in adult life
B. Development of commitment to a career
C. Distancing from parents as a young person
D. Brain damage
E. A good relationship with someone outside the family

Q1.6   In Piaget's stage of precausal logic, a child can:

A. Reason based on observations
B. Accept false explanations
C. Imitate
D. Detach words from the objects they symbolise
E. Reason about quantity

Q1.7   The following are true about language development in children:

A. Rate of language development is linked with intelligence
B. Skinner proposed that acquisition of language is acquisition of a rule system and a code
C. Bruner saw the basis of grammar in preverbal social exchange between mother and child
D. Chomsky demonstrated that children can copy adult grammar
E. Children become proficient in language because of what they hear others say

Q1.8   The following statements are true about adolescence:

A. Absence of adolescent turmoil indicates maladaptation
B. The key task of adolescence is development of personal identity
C. Identity confusion is a source of serious turmoil in normal teenagers
D. Most adolescents identify with their parents' basic moral principles
E. Social phobia has a special relationship with adolescent development

Q1.9   The following statements on genetic factors on development are correct:

A. There is a heritable component in crimes against property
B. There is a strong genetic loading in juvenile delinquency
C. Twin and adoption studies support a strong genetic contribution in petty criminality
D. A neurobiological correlate for harm avoidance has been found
E. The behavioural characteristics of distractibility has a strong genetic component

Q1.10 The following statements are true regarding development of empathy in children:

A. Newborns do not have the ability to respond to the feelings of others
B. True sympathy develops in the second year of life
C. Reflexive emotional resonance starts after the child develops the ability to distinguish between his or her own feelings and those of others
D. The empathic responses become attuned to needs of others from the third year onwards
E. According to Hoffmann, development of empathic responsiveness is dependent on the prospects of reward and punishment by the parents

Q1.11 According to Bentovim, successful families should include:

A. A model for socialisation
B. Models in the parents for sexual identity
C. Mutual dependency and investment
D. Triangular relationship structures
E. Boundaries demarcating parents and children

Q1.12 According to West and Farrington, the following factors predict criminality:

A. Low intelligence
B. Criminality in the father
C. Criminality in the mother
D. Families with more boys than girls
E. Poor parental conduct

Q1.13 Names associated with attachment theory include:

A. Harlow
B. Rutter
C. Lorenz
D. Kliene
E. Winnicot

Q1.14 The following statements are true about the development of antisocial behaviour:

A. Twin and adoption studies have reported a strong genetic influence in juvenile delinquency
B. There is research evidence to show that antisocial behaviour is more closely related to social class than with environmental influence
C. Recent studies have shown a direct relationship between delinquency and family size
D. Family discord is more important than family breakdown as a predictor of future delinquency
E. Multiple regression techniques are better predictors of delinquency than linear methods

Q1.15 The following are true about ethological studies:

A. Tinbergen coined the term *ethology*
B. Imprinting can occur only during a brief period after birth
C. Imprinting affects later sexual behaviour
D. Japanese orphans brought up in China developed Chinese mannerisms
E. Chinese orphans brought up in Japan retained Chinese mannerisms

1. Answers

A1.1
A. T
B. T
C. F It takes 4 years for a child to achieve this milestone
D. T
E. F It takes 4 years for a child to achieve this milestone

A1.2
A. F There is no such time restriction, although this phenomenon has been termed *bonding* by some
B. T
C. F It is not, but many assume that it is
D. T
E. T

A1.3
A. T
B. F Truly intentional social behaviour appears around 3 months
C. T
D. F *Motherese* refers to modification of the mother's speech when interacting with her child
E. F Adult level of vision is reached by 6 months

A1.4
A. F Temperament is an interactional concept between gene and environment
B. F It used the NYLS Parent Questionnaire
C. T
D. F Approach/withdrawal accounted for the greatest variance
E. F Slow to warm up babies are predisposed to developing phobic disorder

A1.5
A. F It was attributed to development of a gift in middle childhood
B. T
C. T The distancing was sometimes geographical
D. F Poor outcome despite low risk factors was found in those with brain damage, depressive illness or schizoid personality
E. T

A1.6

A. F Reasoning is based on the child's internal model of the world
B. T
C. T
D. F A child cannot detach words from the objects they symbolise
E. T

A1.7

A. T
B. T
C. T
D. F They cannot, but invent their own grammatical rules
E. F They become proficient as a result of encouragement to speak

A1.8

A. F It indicates health
B. F Most adolescents are content to adopt the identity ascribed to them by their family of origin
C. F It is a common issue among disturbed adolescents
D. T
E. T

A1.9

A. T
B. F The genetic component is of much lesser magnitude
C. F The genetic contribution in petty criminality is weak
D. T
E. F Distractibility has a weak genetic component

A1.10

A. F Newborns do have the ability to respond to the feelings of others
B. T
C. F Newborns have this capacity
D. T
E. F Is independent of prospects of reward and punishment

A1.11

A. T
B. T
C. F
D. F
E. T

A1.12
A. T
B. T
C. T Criminality in the mother is more predictive than criminality in the father
D. F There is no such relationship
E. T

A1.13
A. T
B. F
C. T
D. F
E. F

A1.14
A. F Only moderate genetic influence has been found in antisocial behaviour
B. F Antisocial behaviour is more closely related to environmental influence than social class
C. F There is only an assumed link
D. T
E. F Multiple regression gives poorer prediction

A1.15
A. F Lorenz did
B. T
C. T
D. F They retained Japanese mannerisms because of the enduring effects of early parent–child interaction with their Japanese parents
E. F No such studies have been done

2. Genetics

Q2.1   The following statements about the human genome are correct:

A. In a normal human genome, about 50% of total genomic DNA is non-coding
B. It is estimated that 30–50% of the human genome is expressed mainly in the brain
C. Mitochondrial chromosomes are identical to nuclear chromosomes
D. During mutation, unstable complementary DNA is produced
E. DNA separates its double helix in a reaction catalysed by reverse transcriptase

Q2.2   The following statements about human DNA are correct:

A. Mammalian DNA is supercoiled around histone proteins
B. tRNA has an anticodon at one end to attach to an mRNA
C. Histone proteins are not transcribed
D. Satellite DNAs are inherited in a non-Mendelian fashion
E. cDNA is produced by reverse transcriptase

Q2.3   The following diseases and their abnormal gene locations are correctly paired:

A. Huntington's disease : 4q 16.3
B. Fragile X syndrome : Xp 27.3
C. Lesch–Nyhan syndrome : Xq 26
D. Wilson's disease : 13q 14
E. Friedreich's ataxia : 9q 12

Q2.4   In disorders with autosomal dominant transmission:

A. The phenotypic trait is present in all individuals carrying the dominant allele
B. When one parent is homozygous, half the offspring will manifest the abnormal trait
C. When a normal individual mates with a heterozygote individual, three quarters of the offspring will manifest the abnormal trait
D. When two heterozygotes mate, all the offspring will manifest the abnormal trait
E. Male and female offspring have equal risk of being affected

Q2.5   In disorders with autosomal recessive transmission:

A. Only double heterozygotes manifest the abnormal phenotypic trait
B. Single heterozygotes are carriers of the abnormal trait
C. When two heterozygotes mate, half the offspring will be affected
D. When an affected individual mates with a normal one, a quarter of the offspring will be affected
E. Male and female offspring have an equal risk of being affected

Q2.6   In disorders with X-linked recessive transmission:

A. Only male offspring will manifest the abnormal phenotypic trait
B. When an affected male mates with a normal female, half the daughters will be carriers
C. When an affected female mates with a normal male, all the daughters will be carriers
D. When a carrier female mates with a normal male, half the daughters will be carriers
E. Half the female heterozygotes are carriers

Q2.7   The following statements are true about twin studies in psychiatry:

A. Monozygotic twins have a more dissimilar environmental influence than dizygotic twins
B. The concordance rate in dizygotic twins is higher than in monozygotic twins when they are born out of wedlock
C. The concordance rate in monozygotic twins is higher than in dizygotic twins in parental assortative mating
D. Pairwise concordance gives lower rates than probandwise concordance
E. Taking a hospitalised sample is the best way to study twin concordance

Q2.8   The following statements are true about adoption studies in psychiatry:

A. In adoptee family studies, illness rates are compared in the biological and adoptive parents of normal adoptees
B. In an adoptee study, illness rates are compared in the biological and adoptive parents of ill adoptees
C. In cross-fostering studies, illness rates are compared in fostered and normal children
D. Most cases for adoption studies are taken from the adoption register
E. The process of adoption is non-random

Q2.9   The following statements are correct:

A. A gene is said to have variable expressivity when it causes varying manifestation of an exophenotype depending on variations in environmental factors
B. Single gene defects are inherited in a non-Mendelian fashion
C. Heritability of a trait is defined by the frequency of expression of a dominant gene
D. Penetrance of a trait is defined as the proportion of total phenotypic variance contributed by the genetic component
E. Characters that are polygenically inherited do not have a continuous population distribution

Q2.10 The following statements are correct:

A. A codon is a sequence of three nucleotide bases that can code for an amino acid
B. In the somatic cell nucleus, each gene exists in an allelic pair
C. During the process of hybridisation, complementary DNA strands are separated
D. A recombinant DNA is derived from more than one organism
E. Cloning involves insertion of a DNA fragment into a cosmid capable of autonomous replication in a host cell

Q2.11 A gene probe:

A. Is a fragment of cDNA
B. Can be constructed by the enzyme restriction endonuclease
C. Can be constructed from mRNA
D. Has a base sequence complementary to that of a given part of a genome
E. Is specifically related to the disease gene for most diseases

Q2.12 Restriction fragment length polymorphism (RFLPs):

A. Are inherited in a simple Mendelian pattern
B. Can be used as DNA markers
C. Detect very short sequences of repeated dinucleotides
D. Are DNA fragments of different lengths in an individual
E. Can be used as gene probes

Q2.13 The following statements about recombination in genetics are correct:

A. The closer two loci are during meiosis, the higher are their chances of recombination
B. Recombinant function measures the frequency of separation of alleles during meiosis
C. The value of recombinant function varies between 0 and 1
D. Linkage measures the frequency of inheritance of two genes in close proximity
E. Lod scores are most effective for conditions with a non-Mendelian mode of inheritance

Q2.14 The following statements about association studies in molecular genetics are true:

A. They are in most respects more complicated than linkage studies
B. They compare the frequency of a particular marker gene in patients and in healthy controls
C. In pleiotropy, different genes apparently affect the same phenotypes
D. In linkage disequilibrium, the recombination fraction is very low
E. Association studies enable detection of genes accounting for a very small proportion of variance in liability

Q2.15 In the liability threshold model of transmission of psychiatric disorder:

A. It is supposed that there is an underlying graded genetic liability to develop the disorder
B. In the general population, plotting the frequency of the disorder against its liability yields a normal distribution
C. In patients suffering from a particular disorder, the curve is shifted to the left
D. The aim of segregation analysis is to find out how much of the environmental influence is due to shared and how much to non-shared influence
E. The sib-pair method aims to elucidate the mode of transmission of a trait rather than simply to estimate its variance component

2. Answers

A2.1
A. **F** 90% of genomic DNA is non-coding
B. **T**
C. **F** There are two copies of nuclear chromosomes in every cell, but there are numerous mitochondrial chromosomes
D. **F** mRNA is produced
E. **F** DNA polymerase catalyses DNA separation

A2.2
A. **T**
B. **T**
C. **T**
D. **F** Satellite DNAs are inherited in a Mendelian fashion
E. **T**

A2.3
A. **F** 4p 16.3
B. **T**
C. **T**
D. **T**
E. **T**

A2.4
A. **T**
B. **F** All offspring would be affected
C. **F** Half will manifest it
D. **F** Three quarters will manifest it
E. **T**

A2.5
A. **F** Only homozygotes manifest it
B. **T**
C. **F** A quarter will be affected
D. **F** None will be affected
E. **T**

A2.6
A. F None of the sons will be affected when an affected male mates with a normal female
B. F All the daughters will be carriers
C. T
D. T
E. F All female heterozygotes are carriers

A2.7
A. T Due to competing effect with the co-twin
B. F
C. F The reverse is true
D. T
E. F Twin concordance is best studied from a twin register, as hospitalised samples contain extremely ill twins, which may give falsely higher concordance rates

A2.8
A. F
B. F
C. F
D. F Adoption registers are not commonly used
E. T An effort is made to find parents similar in character to adoptees

A2.9
A. F Such variations are independent of environmental factors
B. F Single gene defects are inherited in a Mendelian pattern
C. F This is the penetrance
D. F This is the heritability
E. F They do have continuous distribution

A2.10
A. T
B. T
C. F They are brought together; strands are separated in denaturation
D. T
E. T

A2.11
A. T
B. T
C. T By reverse transcriptase
D. T
E. F It is only linked to the disease gene, not specifically related

A2.12
A. T
B. T
C. T
D. F They are DNA fragments of similar lengths
E. F

A2.13
A. F
B. T
C. F It varies between 0 and 0.5
D. T
E. F Lod scores are most effective for conditions with Mendelian inheritance

A2.14
A. F They are easier than linkage studies
B. F They compare marker phenotypes
C. F The same gene affects two or more different phenotypes
D. T Because the disease and marker genes are too close to each other
E. T

A2.15
A. T
B. T
C. F It is shifted to the right
D. F The aim is to elucidate the mode of transmission of a trait rather than simply to estimate its variance component
E. F It is a more robust method of linkage analysis

3. Neurosciences

Q3.1 In cranial nerve palsies:

A. One of the first signs of a compressive pathology in the pituitary region is a bitemporal field defect
B. Progressive failure of gaze in all directions may be associated with Parkinson's syndrome
C. Lesions in the cerebellopontine angle cause loss of corneal sensation
D. Gaze is directed towards the paralysed side in lesions affecting the frontal lobe
E. The disc margins appear raised in hypermetropic eyes on fundoscopy

Q3.2 By making a patient stand with eyes shut and hands out-stretched, one can check:

A. Pyramidal function
B. Extrapyramidal function
C. Parietal function
D. Frontal function
E. Cerebellar function

Q3.3 In the above posture:

A. The palms must face upwards
B. Minimal pyramidal loss causes only spreading of the fingers
C. Drifting of the whole hand with an abnormal posture is a sign of extra-pyramidal disease
D. Cerebellar disease causes the arm to drift upwards and lose position
E. In parietal lesions, the posture of the hand may be abnormal and contin-uously changing

Q3.4 The following are limbic pathways:

A. Stria terminalis
B. Stria medullaris
C. Median forebrain bundle
D. Medial longitudinal stria
E. Dorsal longitudinal stria

Q3.5   The following statements about brain structures are true:

A. The striatum is made up of the caudate nucleus and the globus pallidus
B. The pars reticulata uses dopamine as its main neurotransmitter
C. Basal ganglia output nuclei exert cholinergic inhibition on the thalamus
D. During development, the first brain structure to myelinate is the pre-frontal cortex
E. The whole of the nervous system develops from the ectoderm of the embryo

Q3.6   The following are tests of frontal lobe function:

A. Figure fluency
B. Cognitive estimation
C. Multiple loops test
D. Three step hand sequence
E. Proverb interpretation

Q3.7   The following relationships between brain structures and functions are correct:

A. Destruction of the ventromedial hypothalamus : hyperphagia
B. Limbic system activation : kindling
C. Septohippocampal system : anxiety modulation
D. Amygdala : memory for emotional events
E. Anterior hypothalamic lesion : activation of sexual activity

Q3.8   The following are potential hypothalamic peptide neurotransmitters:

A. Bombesin
B. Bradykinin
C. Secretin
D. Angiotensin
E. α-Endorphin

Q3.9   The following neurotransmitters and peptides coexist:

A. Dopamine : encephalin
B. Noradrenaline : encephalin
C. Acetylcholine : CCK (cholecystokinin)
D. Serotonin : TRH (thyrotropin-releasing hormone)
E. Adrenaline : encephalin

Q3.10 In Huntington's chorea:

A. The brain is normal in size
B. The frontal lobes degenerate
C. GAD levels are low
D. The average age of onset is in the mid-twenties
E. Insight is lost early

Q3.11 The following statements about cerebral tumours are true:

A. Adults suffer mainly from supratentorial tumours
B. Ependymomas spread via the cerebrospinal fluid
C. Meningiomas grow rapidly
D. Medulloblastomas are the commonest primary tumours in childhood
E. Acoustic neuromas affect cranial nerves V, VI, VII and VIII

Q3.12 In Parkinson's disease:

A. Depigmentation is seen mainly in the zona reticulosa
B. Lewy bodies are seen in the dead neurons
C. Cortical atrophy is rare
D. The cranial nerves are spared
E. The reticular formation is affected

Q3.13 The following statements about cerebral inclusion bodies are true:

A. Pick bodies are argyrophilic extracytoplasmic neuronal inclusions
B. Lewy bodies are hyaline intracytoplasmic neuronal inclusions
C. Hirano bodies are eosinophilic intracytoplasmic neuronal inclusions
D. Opalski cells are found in Hunter's syndrome
E. Zebra bodies are found in Hurler's disease

Q3.14 In multi-infarct dementia:

A. The degree of cognitive impairment correlates with the extent of infarction
B. Features of Klüver–Bucy syndrome may be seen
C. In contrast to Alzheimer's disease, the sex ratio is equal
D. A minimum of 100 ml of infarction is required before cognitive impairment is detectable
E. A volume of 100 ml is particularly likely to be associated with dementia

Q3.15 Neurofibrillary tangles are seen in:

A. Amyotrophic lateral sclerosis
B. Down's syndrome
C. Pick's disease
D. Punch-drunk syndrome
E. Dementia of frontal lobe type

Q3.16 In Punch-drunk syndrome:

A. The lateral ventricles are commonly enlarged
B. Cerebral atrophy is unusual
C. Neuritic plaques are sometimes visible
D. Confabulation is a feature found commonly
E. The corpus callosum is perforated

Q3.17 Demyelination occurs in the following conditions:

A. Schilder's disease
B. Niemann–Pick disease
C. Adrenoleucodystrophy
D. Gaucher's disease
E. Tay–Sach's disease

Q3.18 In Wernicke–Korsakoff syndrome:

A. The pathological changes are asymmetrical
B. Demyelination is seen
C. The cerebellum is spared
D. Cerebral atrophy is associated with Wernicke's disease
E. Ventricular dilatation may be seen in Wernicke's disease

Q3.19 Parkinsonism may be caused by:

A. Mercury poisoning
B. Carbon dioxide poisoning
C. Lead poisoning
D. Cerebral palsy
E. Wilson's disease

Q3.20 The new variant of CJD:

A. Presents predominantly with symptoms of dementia
B. Has a course that is slower than that of typical CJD
C. Has scanty PrP amyloid plaques
D. Shows type 4 PrP bands on electrophoresis
E. Has only valine at codon 129

Q3.21 The following statements about evoked potentials are true:

A. The initial waves in an evoked potential tracing are dependent on the psychological state of the individual
B. The P300 wave is said to relate to a process of cognitive appraisal of the stimulus
C. Contingent negative variation (CNV) is another name for 'readiness potential'
D. In schizophrenia, the amplitude of the P300 wave is usually increased
E. The latency of the P300 wave is significantly delayed in dementias

Q3.22 In normal sleep:

A. After intense physical exercise, the proportion of slow wave sleep increases
B. After a lengthy period of wakefulness, the proportion of REM sleep increases
C. Very little stage 2 sleep is regained following a period of sleep deprivation
D. Stage 3 sleep predominates in optional sleep
E. Few sleep spindles are seen in stage 4 sleep

Q3.23 The following statements are true of a normal EEG:

A. Beta and theta rhythms dominate at birth
B. Alpha rhythm is established around 6 years
C. Posterior temporal theta activity is not infrequently seen in younger people
D. Paroxysmal focal spikes are common in children
E. Only a limited number of channels can be recorded in ambulatory EEG

Q3.24 The following statements are true of normal sleep:

A. Delta rhythm is predominant in stage 3
B. It is easier to wake people from REM than from NREM sleep
C. More than 50% of dreams in REM sleep can be remembered
D. There is no evidence to suggest that eye movements are related to dream content during REM sleep
E. The biological effect of total sleep deprivation is severe in the stress-related endocrine system

Q3.25 The following statements are true of EEG:

A. Electrical activity can be measured as early as 12 weeks in the human foetus
B. Immaturity is defined as the presence of an excess of slow waves for the age
C. During ageing, alpha rhythm is better preserved in men
D. Lithium can produce bursts of theta activity
E. Opiates produce little EEG change when taken by addicts

Q3.26 The following statements about neuroreceptors are true:

A. D5 receptors inhibit adenylate cyclase
B. D5 receptors are predominantly seen in the prefrontal cortex
C. D1 receptor activation can enhance intracellular D2 receptor activity
D. $5HT_3$ receptors are linked to adenylate cyclase
E. Stimulation of 5HT receptors decreases acetylcholine release

Q3.27 The following statements about neurophysiological studies of the brain are correct:

A. In MRI studies, T1 relaxation time is always greater than T2 relaxation time
B. T1 relaxation time relates to interaction of proton molecules with each other
C. Image resolution in magnetic resonance spectroscopy is less than in MRI
D. A patient cannot have more than one functional MRI scan at a time
E. SPECT studies require an on-site cyclotron

Q3.28 The following statements about synaptic transmission are correct:

A. The primary structure is similar for all ion channels
B. Amplification is an important feature of the inositol phosphate second messenger system
C. All postsynaptic receptors have five subunits in their general structure
D. The amount of neurotransmitter released during an action potential is related to the calcium levels at the presynaptic terminal
E. Inhibitory postsynaptic potential results from an influx of potassium and chloride

Q3.29 The following statements about prolactin are correct:

A. Prolactin levels tend to be elevated in patients receiving intramuscular therapy
B. Prolactin levels tend to be elevated in patients receiving intravenous therapy
C. Prolactin levels correlate poorly with serum levels of neuroleptics
D. Prolactin levels are predictors of dopamine blockade at the nigrostriatal axis
E. β-Endorphins stimulate prolactin release

Q3.30 The neural crest gives rise to:

A. The dorsal root ganglia
B. The cerebral grey matter
C. The spinal nerves
D. The cranial nerves
E. The adrenal cortex

Q3.31 The following arteries are branches of the internal carotid artery:

A. Posterior choroidal
B. Posterior cerebral
C. Ophthalmic
D. Labyrinthine
E. Anterior choroidal

Q3.32 The following substances pass through the blood–brain barrier by simple diffusion:

A. Carbon dioxide
B. Water
C. Alcohol
D. Glucose
E. Amino acids

Q3.33 The thalamus:

A. Is part of the telencephalon
B. Receives input from all the senses
C. Lateral geniculate nucleus receives information from the ears
D. Anterior nucleus receives taste sensation
E. Dorsomedial nucleus receives general sensation

Q3.34 Infarction in the territory of the anterior cerebral artery leads to:

A. Aphasia
B. Ipsilateral Horner's syndrome
C. Contralateral sensory loss
D. Contralateral hemiplegia
E. Clouding of consciousness

Q3.35 The following types of nystagmus are correctly paired with their causative lesions:

A. Horizontal : middle ear disease
B. Horizontal : cerebellar disease
C. Vertical : multiple sclerosis
D. Vertical : brain stem disease
E. Ataxic : phenytoin overdose

Q3.36 The principal outputs of the basal ganglia go to:

A. The red nucleus
B. The tectum
C. The cerebral cortex
D. The subthalamic nucleus
E. The substantia nigra

Q3.37 Diplopia may occur in:

A. Neuropathy of the optic nerve
B. Parkinson's disease
C. Huntington's chorea
D. Wilson's disease
E. Neuropathy of the oculomotor nerve

Q3.38 Serotonergic cell bodies are found in:

A. The substantia nigra
B. The dorsal raphe nucleus
C. The pontomedullary region of the brain stem
D. The corpus callosum
E. The median raphe nucleus

Q3.39 The following statements are true about glycine as a neurotransmitter:

A. It is predominantly found in the brain
B. Its principal effect is on the presynaptic membrane
C. It is a weak excitatory neurotransmitter
D. It hypopolarises motor neurons
E. Strychnine competes for its receptors

Q3.40 In the brain:

A. The lipid content is minimal
B. Gangliosides are the predominant type of lipid
C. Less than 1% of cholesterol is in the free form
D. The brain lipids are relatively unaffected by dietary lipids
E. Triglycerides and free fatty acids are abundant

Q3.41 The following statements regarding substance P are correct:

A. It consists of 11 amino acids
B. A high concentration is found in the cortex
C. There is a large nigrostriatal pathway involving it in the brain
D. It is most likely to be an inhibitory neurotransmitter
E. It can cause both hyperalgesia and analgesia

Q3.42 The following receptors are cation channel linked:

A. Nicotinic
B. Glycine
C. $5HT_3$
D. $5HT_2$
E. Delta

Q3.43 The following receptors are G protein linked:

A. GABA-B
B. Dopamine D1 and D2
C. $5HT_{1C}$
D. $5HT_2$
E. Muscarinic

Q3.44 The following components of language are correctly related to their anatomical structures:

A. Phonology : left superior temporal lobe
B. Semantics : left temporal lobe
C. Syntax : left anterior hemisphere
D. Prosody (fine tuning) – right hemisphere
E. Prosody (emotional expression) – left anterior hemisphere

Q3.45 The following neuropathological features of Parkinson's disease are correct:

A. Pallor of the locus ceruleus
B. Pallor of Meynert's nucleus
C. Pallor of the Edinger–Westphal nucleus
D. Pallor of the sympathetic nuclei
E. Pallor of the vagus nucleus

Q3.46 The following inhibit growth hormone secretion:

A. Increased free fatty acids
B. Obesity
C. Hepatic cirrhosis
D. Insomnia
E. Renal failure

Q3.47 The following are found in sociopaths:

A. Increased cortical arousal
B. Decreased skin conductance
C. Exaggerated response to stress
D. Rapid development of conditioning to fear-provoking stimuli
E. Diffuse slow wave on EEG

Q3.48 The following stimulate prolactin secretion:

A. Cushing's disease
B. Renal failure
C. Empty sella syndrome
D. Secondary hypothyroidism
E. Hepatic cirrhosis

Q3.49 The following signs are correctly paired with their site of origin:

A. Anosognosia : dominant parietal
B. Prosopagnosia : non-dominant parietal
C. Alexia with agraphia : dominant temporal lobe
D. Complex visual hallucination : non-dominant occipital lobe
E. Astereognosis in the right hand : corpus callosum

Q3.50 Hyperprolactinaemia is associated with:

A. Paedophilia
B. Ejaculatory failure
C. Erectile failure
D. Klinefelter's syndrome
E. Libidinal failure

Q3.51 The following statements are true about the GABA shunt:

A. It is unique to the brain
B. It accounts for a quarter of the total glucose turnover in the brain
C. It is a bypass around the tricarboxylic acid cycle from ketoglutarate to succinate
D. It arises because of the presence of the enzyme glutamate transaminase
E. It requires vitamin B6 as a cofactor

Q3.52 The following statements are true about biochemical markers:

A. DNA can be used as an index of cell numbers
B. Cholesterol can be used as an indicator of myelin
C. DNA can be used as an indicator of cell size
D. Acetylcholinesterase can be used as an indicator of synapses
E. Gangliosides can be used as indicators of glial cells

Q3.53 SPECT:

A. Employs radiochemicals that emit positrons
B. Has better image resolution in than positron emission tomography (PET)
C. Can measure metabolism in different areas of brain in one individual
D. Does not have any radiation risk
E. May be used to compare cerebral metabolic measurements within patient groups

3. Answers

A3.1
A. **F** Bitemporal loss of red colour appreciation is one of the first signs
B. **T** When associated with progressive supranuclear palsy
C. **T**
D. **F** It is directed towards the side of the lesion
E. **T**

A3.2
A. **T**
B. **F**
C. **T**
D. **F**
E. **T**

A3.3
A. **F** They may face downwards
B. **T**
C. **F** This indicates severe pyramidal disease
D. **F** Parietal disease causes this
E. **F** Cerebellar disease causes this

A3.4
A. **T**
B. **T**
C. **F** The medial forebrain bundle is a limbic pathway
D. **T**
E. **F** The dorsal longitudinal *fasciculus* is a limbic pathway

A3.5
A. **F** It consists of the caudate and the putamen
B. **F** Its main neurotransmitter is GABA; the pars compacta uses dopamine
C. **F** They exert GABAergic inhibition
D. **F** This is the last structure to myelinate
E. **T**

A3.6
A. T
B. T
C. T
D. T
E. T

A3.7
A. T
B. T
C. T
D. T
E. F Anterior hypothalamic lesions are associated with prevention of sexual activity

A3.8
A. T
B. T
C. T
D. T
E. F β-endorphins are potential neurotransmitters

A3.9
A. T
B. T
C. F Acetylcholine and VIP (vasoactive intestinal polypeptide) coexist
D. T
E. T

A3.10
A. F The brain is small in size
B. F
C. T
D. F Average onset is in the thirties
E. F Insight is retained

A3.11
A. T
B. T
C. F Meningiomas are usually slow growing
D. T
E. T

A3.12
A. T
B. F Lewy bodies are seen in surviving neurons
C. F Diffuse cortical atrophy occurs
D. F The dorsal vagal nerve nucleus is affected
E. T

A3.13
A. F Pick bodies are intracytoplasmic
B. T
C. T
D. F Opalski cells are seen in Wilson's disease
E. T

A3.14
A. T
B. F These are seen in Alzheimer's disease
C. F Males are more commonly affected
D. F A minimum of 50 ml of infarcted brain tissue is required before cognitive impairment is detected
E. T

A3.15
A. T
B. T
C. F
D. T
E. F

A3.16
A. T
B. F Gross atrophy occurs
C. F Plaques are not seen
D. F Confabulation is not found
E. F The septum pellucidum is perforated

A3.17
A. T
B. T
C. T
D. T
E. T

A3.18

A. F The changes are bilaterally symmetrical
B. T
C. F The superior vermis of the cerebellum is affected
D. F Atrophy is seen only in Korsakoff's syndrome
E. T

A3.19

A. T
B. F Carbon monoxide poisoning can cause parkinsonism
C. T
D. T
E. T

A3.20

A. F New variants of CJD present predominantly with psychiatric symptoms
B. T
C. F Numerous PrP amyloid plaques are seen
D. T
E. F All the amino acids at codon 129 are methionine

A3.21

A. F They are independent of psychological state
B. T
C. F Readiness potential is a positive motor potential arising 1 second before voluntary movement
D. F The amplitude of the P300 wave is decreased in schizophrenia
E. T

A3.22

A. T
B. F Slow wave sleep increases after lengthy wakefulness
C. T
D. F Stage 2 sleep predominates in optional sleep
E. F Sleep spindles are no longer seen in stage 4 sleep

A3.23

A. F Delta rhythm dominates at birth
B. F Alpha rhythm is established around 13 years
C. T
D. F They are seen in only 3% of children
E. T

A3.24

A. F Delta sleep is predominantly seen in stage 4
B. F The opposite is true
C. T
D. F Some evidence is emerging
E. F Surprisingly, it is extremely limited

A3.25

A. F Electrical activity arises at 20 weeks in a human foetus
B. T
C. F It is better preserved in women
D. F Lithium produces delta rhythm, which may be focal
E. T

A3.26

A. F D5 receptors stimulate adenylate cyclase
B. F D4 receptors are predominantly seen in the prefrontal cortex
C. T
D. F $5HT_3$ receptors are linked to ion channels
E. T

A3.27

A. T
B. F T1 relaxation time relates to interaction of protons with surrounding nuclei
C. T
D. F Multiple scans are possible as it is non-invasive and does not require radioactive substances
E. F They do not, as the tracers used have a long half-life

A3.28

A. T
B. F Amplification is an important feature of the adenylate cyclase system because each activated receptor protein stimulates many G protein molecules, which in turn activate many molecules of adenylate cyclase, each generating many cAMP molecules
C. T
D. T
E. T

A3.29

A. T

B. F

C. T

D. F Prolactin levels predict dopamine blockade at the hypothalamo-pituitary axis

E. T

A3.30

A. T

B. F The neural tube gives rise to the cerebral grey matter

C. T

D. T

E. F The neural crest gives rise to the adrenal medulla

A3.31

A. F This is a branch of the posterior cerebral artery

B. F This is a branch of the basilar artery

C. T

D. F This is a branch of the basilar artery

E. T

A3.32

A. T

B. T

C. T

D. F

E. F Glucose and amino acid need facilitated transport

A3.33

A. F The thalamus is part of the diencephalon

B. F It receives no input from the sense of smell

C. F It receives information from the eyes

D. F The ventral posterior nucleus receives taste

E. F The ventral posterior nucleus receives general sensation

A3.34

A. T

B. T

C. T

D. T

E. T

A3.35
A. T
B. T
C. F
D. T
E. F   Ataxic nystagmus is highly suggestive of multiple sclerosis

A3.36
A. T
B. T
C. F
D. T
E. T

A3.37
A. F   It occurs in neuropathy of the oculomotor nerve
B. F
C. F
D. F
E. T

A3.38
A. T
B. T
C. T
D. F
E. T

A3.39
A. F   Glycine is predominantly found in the spinal cord
B. F   Its principal effect is on the postsynaptic membrane
C. F   It is a potent inhibitory neurotransmitter
D. F   It hyperpolarises postsynaptic membrane
E. T

A3.40
A. F   The brain is rich in lipids
B. F   Cholesterol is the predominant lipid in the brain
C. F   Most of the cholesterol exists in the free form
D. T
E. F   Very little is found in the brain

A3.41
A. T
B. F High levels are found in the basal ganglia and hypothalamus
C. T
D. F It is an excitatory neurotransmitter in pain signalling and for certain dopaminergic pathways
E. T

A3.42
A. T These receptors are sodium channel linked
B. F These receptors are chloride channel linked
C. T These receptors are sodium channel linked
D. F These receptors are phosphoinositol linked
E. T These receptors are potassium channel linked

A3.43
A. T
B. T
C. F These receptors are phosphoinositol linked
D. F These receptors are phosphoinositol linked
E. T

A3.44
A. T
B. T
C. T
D. F Left anterior hemisphere
E. F Right hemisphere

A3.45
A. T
B. T
C. F
D. T
E. T Pallor of the dorsal motor nucleus is a feature

A3.46
A. T
B. T
C. F Hepatic cirrhosis stimulates growth hormone secretion
D. T
E. F Renal failure stimulates growth hormone secretion

A3.47
A. **F** Cortical arousal is decreased
B. **T**
C. **F** There is limited response to stress
D. **F** Conditioning to fear-provoking stimuli occurs slowly
E. **T**

A3.48
A. **T**
B. **T**
C. **T**
D. **F** Primary hypopituitarism stimulates prolactin secretion
E. **F** There is no association

A3.49
A. **F** The non-dominant parietal lobe is involved
B. **T**
C. **T**
D. **T**
E. **F** Astereognosis in the left hand is caused by lesions in the corpus callosum

A3.50
A. **T**
B. **T**
C. **T**
D. **T**
E. **T**

A3.51
A. **T**
B. **F** About 10% of total glucose turnover in the brain occurs via the GABA shunt
C. **T**
D. **F** It requires glutamic acid decarboxylase
E. **T**

A3.52
A. **T**
B. **T**
C. **F** DNA can be used as an indicator of protein/DNA ratio
D. **T**
E. **F** Gangliosides can be used as indicators of brain specific protein S-100

A3.53

A. F SPECT (single photon emission computed tomography) employs radiochemicals that emit single photons

B. F The reverse is true

C. T

D. F There are small radiation risks

E. F This is not possible in SPECT; comparison is possible in PET

4. Psychopharmacology

Q4.1   The following are true about class I drug receptors:

A. They act via ion channels
B. They are slow acting
C. They act via the alpha unit of G protein
D. Muscarinic receptors are good examples of class I drug receptors
E. Nicotinic receptors are good examples of class I drug receptors

Q4.2   The following combinations of drugs are dangerous:

A. Fluoxetine and clomipramine
B. Buspirone and tranylcypromine
C. Lithium and fluvoxamine
D. Triazolam and trimipramine
E. Phenelzine and carbamazepine

Q4.3   Zopiclone:

A. Is a diazolobenzodiazepine
B. Binds to GABA receptors
C. Can be used in lactating mothers
D. Is effective in complex partial seizure
E. Does not suppress REM sleep

Q4.4   Lithium:

A. Inhibits extracellular phosphatase
B. Is indicated in refractory anxiety states
C. Increases adenylate cyclase activity
D. Levels are increased by carbonic anhydrase
E. Shows little variation in bioavailability with different preparations

Q4.5   For a drug whose elimination is first order:

A. The half-life increases as the dose administered increases
B. The elimination rate is constant
C. Linear kinetics is obeyed
D. The steady-state plasma level is proportional to the dose
E. The half-life is proportional to the plasma concentration

Q4.6 Following oral administration, drugs are absorbed:

A. Primarily by active transport
B. Mainly in the stomach
C. Better in the ionised form
D. Less readily in the presence of food
E. More slowly than when given by intramuscular injection

Q4.7 The following are true about opiates:

A. Opiate receptors are found in the thalamus
B. Opiates decrease respiratory depth
C. Opiates decrease the sensitivity of the respiratory centre to carbon dioxide
D. Opiates can be used as an antiemetic
E. Opiates decrease the respiratory rate

Q4.8 The following pairings of receptors and their antagonists are correct:

A. NMDA : dizocilpine
B. M1: pirenzepine
C. Alpha1: clonidine
D. Dopamine1: lisuride
E. $5HT_3$: raclopride

Q4.9 The following combinations of receptors and their agonists are correct:

A. $5HT_{1A}$: Ondansetrone
B. $5HT_{1B}$: RU-24969
C. $5HT_2$: ritanserin
D. Dopamine1 : SKF-38393
E. Dopamine2 : pergolide

Q4.10 The following combinations of drugs and mechanisms of action are correct:

A. Cocaine : inhibits noradrenaline (NA) reuptake
B. Cocaine : inhibits dopamine (DA) reuptake
C. Amphetamine : increases NA release
D. Amphetamine : increases DA release
E. Benztropine : inhibits DA reuptake

Q4.11 The following drugs are correctly paired with adverse effects on the foetus:

A. Diazepam : irritability
B. Diamorphine : floppy baby syndrome
C. Tricyclic antidepressants : tachycardia
D. Phenothiazine : congenital malformation
E. Lithium : Fallot's tetralogy

Q4.12 The following adverse effects result when the drugs are taken with disulphiram:

A. Antidepressants : decreased plasma concentration
B. Phenytoin : psychosis
C. Metronidazole : increased toxicity
D. Benzodiazepine : decreased sedation
E. Anticoagulants : increased bleeding tendencies

Q4.13 The following adverse effects result when the drugs are taken with monamine oxidase inhibitors (MAOIs):

A. Insulin : increased chances of hypoglycaemia
B. Oral antidiabetics : decreased effectivity
C. Anticonvulsants : decreased seizure threshold
D. Pethidine : hypertension
E. Tryptophan : confusion

Q4.14 The following adverse effects result when the drugs are taken with specific serotonin reuptake inhibitors (SSRIs):

A. Beta blockers : increased concentration
B. Clozapine : decreased concentration
C. Haloperidol : increased concentration
D. Tricyclic antidepressants : decreased concentration
E. Carbamazepine : increased concentration

Q4.15 The following are adverse effects of antiandrogenic drugs:

A. Fatigue
B. Breathlessness
C. Depression
D. Thrombosis
E. Hirsutism

4. Answers

A4.1
A. **T**
B. **F** Class I receptors are fast acting
C. **F** This is true for class II receptors
D. **F**
E. **T**

A4.2
A. **T** This combination carries an increased risk of serotonin syndrome
B. **T** Can cause hypertension
C. **F**
D. **F**
E. **T** Can lower seizure threshold

A4.3
A. **F** It is a cyclopyrrolone
B. **T**
C. **F** It is excreted in breast milk
D. **F**
E. **F** It does, but less than benzodiazepines

A4.4
A. **F** Lithium inhibits intracellular phosphatase
B. **F**
C. **F** It inhibits adenylate cyclase
D. **F** Lithium levels are unaffected
E. **F**

A4.5
A. **F** The half-life remains constant
B. **F** The elimination rate is proportional to the plasma concentration
C. **T**
D. **T**
E. **F** The half-life remains constant

A4.6

A. **F** Most drugs are passively absorbed
B. **F** They are absorbed mainly in the small intestine
C. **F** They are absorbed better in the non-ionised form
D. **F** The presence of food increases absorption of most psychotropic drugs
E. **F** Except for certain benzodiazepines, e.g. diazepam

A4.7

A. **T**
B. **F** Opiates decrease the rate, not the depth, of respiration
C. **T** But not to hypoxia
D. **F** They may cause vomiting
E. **T**

A4.8

A. **T**
B. **T**
C. **F** Clonidine is an agonist
D. **T**
E. **T**

A4.9

A. **F** It is an antagonist at $5HT_3$
B. **T**
C. **F** It is an antagonist at $5HT_2$
D. **T**
E. **T**

A4.10

A. **T**
B. **T**
C. **T**
D. **T**
E. **T**

A4.11

A. **F** It causes floppy baby syndrome
B. **F** It causes low birth weight
C. **T**
D. **T**
E. **F** It may cause Ebstein's anomaly

A4.12
A. **F** Increased plasma concentration results
B. **F** This combination causes increased toxicity
C. **F** This combination causes psychosis
D. **F** Increased sedation results
E. **T**

A4.13
A. **T**
B. **T**
C. **T**
D. **F** Hypotension results
E. **T**

A4.14
A. **T** Especially with fluvoxamine
B. **F** Increased concentration results
C. **T**
D. **F** Increased concentration results
E. **T**

A4.15
A. **T**
B. **T**
C. **F** Severe depression is a contraindication
D. **F** It is a contraindication
E. **F** It is an indication for use

5. Psychology

Q5.1 The following statements about the processes of learning of new behaviour are true:

A. Habituation is a complex form of learning
B. Conditioned stimulus is usually of biological significance
C. In classical conditioning subjects are passive
D. The main effect of operant conditioning is to increase the number of different stimuli to elicit a given response
E. In classical conditioning new behaviour can be learnt

Q5.2 The following are correctly paired:

A. Tolman : insight learning
B. Köhler : sign learning
C. Bandura : social learning
D. Thorndike : law of mass
E. Skinner : law of effect

Q5.3 Models from whom observational learning takes place have:

A. High status
B. High competence
C. High social power
D. High speed while speaking
E. Higher attractiveness

Q5.4 The following statements about different memory processes are correct:

A. Recency effect means that the first words learned are retained better
B. Episodic memory is subject to effort after meaning
C. Ribot's law states that retrograde amnesia affects more remote memories
D. The forgetting curve has a sharp initial gradient
E. The encoding specificity principle states that in certain cases recognition can be better than recall

Q5.5   Perceptual constancy has been demonstrated for:

A. Lightness
B. Location
C. Height
D. Colour
E. Depth

Q5.6   Perceptual set can be affected by:

A. Hunger
B. Thirst
C. Punishment
D. Personality
E. Experience

Q5.7   The following statements about perceptual threshold are correct:

A. Absolute threshold is tested by the method of descending limits
B. Large differences between two stimuli can be detected at low intensities, but only small differences can be detected at high intensities
C. The Weber–Fechner law measures the lowest intensity of a detectable stimulus
D. The Weber–Fechner law measures the smallest detectable change between two stimuli
E. The relationship between threshold of a stimulus and intensity is an inverted U shaped

Q5.8   The following statements about perceptual threshold are correct:

A. Absolute threshold is taken as the minimum amount of energy required to activate half the sensory organs
B. Difference threshold is the same as 'just noticeable difference'
C. Weber's law holds over a large range of stimulus intensities
D. Fechner's law states that sensory perception is a logarithmic function of stimulus intensities
E. Weber's constant for brightness of light is $\frac{1}{60}$

Q5.9   The following statements about learning theories are correct:

A. Theoretically, it is impossible to learn while under the influence of drugs
B. Cognitive learning is a type of associative learning
C. Delayed conditioning is optimal when the delay between two stimuli is about 0.5 second
D. Simultaneous conditioning is superior to delayed conditioning
E. In trace conditioning, the conditioned stimulus terminates before onset of the unconditioned stimulus

Q5.10 The following statements about learning theories are correct:

A. Thorndike is associated with operant behaviour
B. Skinner is associated with respondent behaviour
C. Operant behaviour is independent of stimulus
D. Respondent behaviour is independent of stimulus
E. Unconditioned responses in classical conditioning are a type of respondent behaviour

Q5.11 The following statements about operant conditioning are true:

A. Discrimination can occur in operant conditioning
B. The phenomenon of extinction is restricted to classical conditioning
C. A negative reinforcer reduces the probability of occurrence of an operant behaviour
D. Punishment is a type of negative reinforcement
E. In partial reinforcement, all the conditioned responses are partially reinforced

Q5.12 The following statements about the applications of associative learning are true:

A. Reciprocal inhibition is a psychological concept
B. Deep muscular relaxation is the most important element of systematic desensitisation
C. Exposure to the feared stimulus is the most important element of systematic desensitisation
D. A proper hierarchical presentation is the most important element of systematic desensitisation
E. Biofeedback is an application of classical conditioning

Q5.13 The following statements are true of non-associative learning:

A. Cognitive learning may occur suddenly
B. Insight learning is an active form of learning
C. In latent learning, learning is not manifested except in an emergency
D. Observational learning is a type of social learning
E. Social learning includes learning by classical conditioning

Q5.14 The following are true of human memory:

A. Olfactory information is stored as haptic memory
B. Short-term memory registers can be filled by parallel processing
C. 75% of information in short-term memory is forgotten by 9 seconds
D. The primacy effect does not hold true if more than seven items are presented for remembering
E. Verbal information in long-term memory is stored as words

Q5.15 The following are correctly paired:

A. Chaining : a method of successive approximation
B. Eysenck : preparedness
C. Mowrer : three-stage theory of phobia
D. Rachman : social skills training
E. Rutter : locus of control

Q5.16 The following statements about motivational theories are correct:

A. Multiple avoidance–avoidance conflicts are the most difficult to resolve
B. Love and belonging needs rank higher than self-esteem needs in Maslow's hierarchy
C. Intrinsic motivation theories suggest that needs arise to maintain biological homeostasis
D. Primary drives are always biological in nature
E. Achievement needs are concerned with the need to have impact, reputation and influence

Q5.17 The following statements regarding personality theories are correct:

A. In Kelly's theory, an individual is perceived as a philosopher
B. Eysenck's theory is a nomothetic type theory
C. Repertory grids can predict future behaviour
D. Roger's self theory suggests that there should be some incongruence between an individual's ideal and real self
E. Ideographic personality theories are based on a psychometric approach

Q5.18 The following statements are correct:

A. Variable interval reinforcement schedules take the longest to establish
B. The concept of incubation explains why some stimuli are more likely to condition than others
C. Perceptual constancy is fully developed by the age of 5 years
D. In competitive situations, an individual with an external locus of control does better than an individual with an internal locus of control
E. The concept of preparedness explains why phobic behaviour increases in severity

Q5.19 The following are correctly paired:

A. Perls : Gestalt therapy
B. Janov : primal therapy
C. Frankl : logotherapy
D. Binswanger : existential analysis
E. Sullivan : interpersonal psychotherapy

Q5.20 There is an inverse correlation between intelligence and:

A. Birth order
B. Family size
C. Age
D. Parental IQ
E. Marital status

Q5.21 The following associations are correct:

A. Spearman : S factor
B. Cattell : primary mental abilities
C. Hebb : type C intelligence
D. Thurstone : fluid ability
E. Vernon : type A intelligence

Q5.22 The following statements about memory types are correct:

A. Episodic and semantic are types of declarative memory
B. Semantic memory is essentially autobiographical in nature
C. In certain circumstances, it is easier to recall than to recognise information
D. Many amnesic patients have poor procedural but intact declarative memory
E. Traditional memory tests measure explicit memory

Q5.23 The following are correctly paired:

A. Adler : analytic psychology
B. Jung : individual psychology
C. Kohut : self psychology
D. Hartmann : ego psychology
E. Horney : basic need

Q5.24 The following statements about self-concept are true:

A. Physical attractiveness plays a prominent role in determination of self-esteem in children
B. Being popular is important in determination of self-esteem, especially in males
C. High self-esteem is associated with more risk-taking behaviour
D. Self-esteem is evaluative in nature
E. Self-image influences behaviour in a systematic and predictive manner

Q5.25 The following statements about stress responses are correct:

A. According to Seyle, most stress responses are non-specific
B. The level of body resistance is increased during the alarm reaction phase
C. Mineralocorticoids are released during the acute stress reaction phase
D. Platelet aggregation is reduced during the acute physiological stress reaction
E. Inflammatory responses are facilitated during acute physiological stress

5. Answers

A5.1
A. F It is the simplest form of learning
B. F Unconditioned stimulus is of biological significance
C. T
D. F This is true for classical conditioning
E. F This is true for operant conditioning

A5.2
A. F Tolman is associated with sign learning
B. F Köhler is associated with insight learning
C. T
D. F Thorndike is associated with law of effect
E. F New behaviours are learnt through operant conditioning

A5.3
A. T
B. T
C. T
D. F There is no association between speed of talking and observational learning
E. F There is no association between the attractiveness of the model and observational learning

A5.4
A. F This is primacy effect
B. T
C. F It affects more recent memories
D. T
E. F Normally recognition is better than recall but, in circumstances where recognition and retrieval contexts match, recall is better than recognition

A5.5
A. T
B. T
C. T
D. T
E. T

A5.6
A. **T**
B. **T**
C. **T**
D. **T**
E. **T**

A5.7
A. **F** Perceptual threshold is measured by the method of ascending limits
B. **F** It's the other way round
C. **F** It states that the relationship between perceptual threshold and intensity is a logarithmic one
D. **F**
E. **F** The relationship is logarithmic

A5.8
A. **F** Absolute threshold is the minimum energy required to activate a sensory organ in 50% of trials
B. **T**
C. **F** It fails to hold over a large range of stimulus intensity
D. **T**
E. **T**

A5.9
A. **T**
B. **F** Classical and operant conditioning are the two forms of associative learning; cognitive learning is a more complex process
C. **T**
D. **F** Delayed conditioning is superior
E. **T**

A5.10
A. **T**
B. **T**
C. **T**
D. **F** It is dependent on known stimuli
E. **T**

A5.11

A. T

B. F It can occur in respondent conditioning as well

C. F It is an aversive stimulus, the removal of which increases the probability of occurrence of an operant behaviour

D. F

E. F Only some of the responses are reinforced

A5.12

A. F It is a neurological concept

B. F

C. T

D. F It is not essential

E. F It is an application of operant conditioning

A5.13

A. T

B. F It occurs suddenly

C. F Learning occurs in response to the need to satisfy a basic drive

D. F It is a type of cognitive learning

E. T

A5.14

A. F Information from touch is stored as haptic memory

B. F Short-term memory registers are filled by the displacement principle

C. T

D. F It can hold true

E. F Verbal information is stored as meanings

A5.15

A. F It is a method of shaping

B. F Seligman is associated with preparedness

C. F Mowrer's is a two-stage theory

D. F Rachman is associated with the three-stage theory of phobia

E. T

A5.16

A. F Multiple approach–avoidance conflicts are the most difficult to resolve

B. F Self-esteem ranks higher than love and belonging needs

C. F This is true for extrinsic motivation theories

D. T

E. F Achievement needs are concerned with need to improve performance, to do better, etc.

A5.17
A. **F** In Kelly's theory, an individual is perceived as a scientist
B. **T**
C. **T**
D. **F** Roger's self theory suggests that there should be some congruence between real and ideal self
E. **F** Nomothetic theories are based on a psychometric approach

A5.18
A. **T**
B. **F** It explains why phobic behaviour increases in severity
C. **T** In some books 6–7 years is quoted as the limit
D. **F** Those with an internal locus of control do better
E. **F** It explains why some stimuli are more likely to condition than others

A5.19
A. **T**
B. **T**
C. **T**
D. **T**
E. **T**

A5.20
A. **T**
B. **T**
C. **F** Intelligence increases with age up to a certain limit
D. **F**
E. **F** There is no such relation

A5.21
A. **T**
B. **F** This is Thurstone's concept
C. **F** This is Vernon's concept
D. **F** This is Cattel's concept
E. **F** This is Hebb's concept

A5.22
A. **T**
B. **F** Episodic memory is autobiographical in nature
C. **T** According to Tulvig's encoding specificity principle
D. **F** Many amnesic patients have better procedural memory
E. **T**

A5.23
A. F Adler should be paired with individual psychology
B. F Jung should be paired with analytic psychology
C. T
D. T
E. T

A5.24
A. T
B. F It is more important in females
C. T
D. T
E. T

A5.25
A. T
B. F Decreased
C. T
D. F It is increased
E. F They are suppressed

6. Psychometry and Research Methodology

Q6.1 The following statements about IQ tests are correct:

A. The Stanford Binet and Wechsler scales correlate poorly
B. The upper IQ limits in Stanford Binet and Wechsler scales are the same
C. Mental age gives a 68% chance of correctly answering a question for the corresponding chronological age
D. Correlations between scores on Raven's Progressive Matrices and the Wechsler Adult intelligence scale (WAIS) are usually of the order of 0.75
E. The Raven's Progressive Matrices test does not involve recall of any learned information

Q6.2 Comparable IQ tests have shown that:

A. Boys are better at skills involving spatial relations
B. Girls are better at mathematics
C. Girls have higher IQ scores in childhood
D. Girls excel in vocabulary
E. Boys are more gifted

Q6.3 The following statements are correct:

A. Studies have shown that 50% of the variance in IQ of offspring is directly related to parental IQ
B. Modern IQ tests are remarkably exact at measuring intelligence
C. There is an inverse relationship between a child's IQ and the socio-economic status of the family
D. Childhood IQ is better at predicting adult educational attainment than adult IQ is
E. There is evidence that boys have a greater range of IQ than girls

Q6.4 The following are 'hold' tests:

A. Comprehension
B. Substitution
C. Vocabulary
D. Similarities
E. Digit span

Q6.5   The following are true about IQ tests:

A. IQ tests can specifically diagnose brain damage
B. In psychiatric illnesses in general, performance on verbal tests is less impaired than on non-verbal tests
C. Performance is better on verbal tests than on non-verbal tests in organic illness
D. Brain damage is more devastating in adults than in children
E. Old people are handicapped by reduction of 'fluid' intelligence

Q6.6   The following are true of personality tests:

A. The Q-sort technique compares the unique configuration of personality traits of different individuals
B. High scorers on the psychotism scale of the Eysenck Personality Questionnaire (EPQ) are claimed to resemble stereotyped psychopaths
C. The Minnesota Multiphasic Personality Inventory (MMPI) can measure defensiveness during answering the questionnaire
D. The EPQ incorporates a lie scale
E. The MMPI measures traits that are part of normal personality

Q6.7   The following statements about IQ tests are correct:

A. Mental ability in the Stanford Binet test is measured by mathematical and problem-solving abilities
B. In the Stanford Binet test, ten items are allocated to each year
C. The highest attainable score of chronological age in the Stanford Binet test is 15
D. Wechsler intelligence scales cannot measure the IQ of children below 5 years of age
E. DSM IIIR allows a measurement error of 10 points on IQ

Q6.8   The following are true statements about rating scales:

A. Forced choice techniques help to eliminate extreme forms of responding
B. Socially desirable responses occur only consciously
C. The phenomenon of social desirability occurs more commonly during interviews than in self-administered questionnaires
D. Lie scales detect deliberate liars
E. Response set occurs when subjects do not want to give away too much self-related information

Q6.9 The following statements about 'self' scales are correct:

A. Self-prediction is a valuable way of measuring behaviour
B. Self-prediction is most useful in alcoholics who are trying to reduce their drinking
C. Self-prediction is very useful to predict reoffending in prisoners
D. Self-recording is a better way of monitoring behaviour than self-prediction
E. Self-prediction is as useful as any other objective method of predicting relapse in smokers

Q6.10 The following statements about behavioural research are correct:

A. The halo effect is a common error made by subjects during interview
B. Video cameras should not be used during naturalistic observations
C. Naturalistic observations may be contaminated by Hawthorne effect
D. Time sampling is a type of naturalistic observation
E. Naturalistic observations can be used to carry out functional analysis of behaviour

Q6.11 The following statements about rating scales are correct:

A. The Likert scale is an equal interval scale
B. The reliability of an equal interval scale improves steadily up to a maximum of 5 points
C. When an investigator is measuring changes in the quality of an interpersonal relationship, it is important to use a generalisable scale
D. It is better to use an interval scale than a categorical scale when studying a psychological phenomenon in detail
E. A highly reliable scale may have poor validity, but a highly unreliable scale may be valid

Q6.12 The following statements about psychological research are true:

A. In any experiment, the more homogeneous a study sample is, the more generalisable the results are likely to be
B. None of the psychophysiological measures are unequivocally true
C. The eyeball approach is a useful tool to measure eye movement sensitivity in post-traumatic stress disorders
D. A rating scale records particular characteristics quantitatively
E. A rating scale can be called a questionnaire when statements are made in a question-and-answer form

Q6.13   The following statements about research methods are correct:

A. Data massaging in an experiment reduces the chances of type I errors
B. When many different raters are being assessed to measure reliability of an instrument, intra-rater reliability is more useful than inter-rater reliability
C. Product moment correlation coefficient is the most frequently used statistic to measure agreement between two raters
D. Product moment correlation coefficient is better than Kappa statistics in measuring the extent of agreement between two raters
E. Kappa statistics give a lower level of agreement than simple correlation measures

Q6.14   The following statements about hypothesis testing are correct:

A. Research in psychiatry is based primarily on proving rather than disproving a hypothesis
B. When a null hypothesis is disproved, it is automatically assumed that a directional hypothesis is proved
C. Recent evidence suggests that calculating an overall effect size for a hypothesis is probably better than doing costly time-consuming prospective studies
D. A pilot study is a small study to test the null hypothesis
E. A cohort study is especially useful in disorders for which no satisfactory treatment is available

Q6.15   In an epidemiological study:

A. A relative risk of 1 implies no causation
B. An attributable risk of 1 implies no causation
C. The relative risk can be calculated only in prospective studies
D. When a disease is relatively rare, an odds ratio can be calculated to check the relative risk
E. A positive attributable risk implies causation

Q6.16   The following statements are true of rating scales:

A. The General Health Questionnaire (GHQ) is particularly useful in identifying somatisation disorder
B. The Schedule for Affective Disorders and Schizophrenia (SADS) is a semi-structured interview schedule
C. The Present State Examination (PSE) can be used to determine caseness
D. SADS is more reliable than PSE in identifying schizophrenia
E. Case registers are useful for conducting cross-sectional studies

Q6.17 The following statements about caseness are correct:

A. Specificity of test is an index for caseness
B. Increasing the caseness threshold of a test increases its sensitivity
C. Increasing caseness threshold of a test decreases its specificity
D. The efficiency of a test is measured by the proportion of all the true results
E. To count as a case, standardised criteria require all the individual component thresholds to be passed

Q6.18 The following statements are true about distribution curves:

A. The arithmetic mean is a good measure of central tendency in a skewed distribution
B. The mode is greater than the mean in a positively skewed distribution
C. The mean is greater than the median in a positively skewed distribution
D. The mean is less than the mode in a negatively skewed distribution
E. The median is greater than the mode in a negatively skewed distribution

Q6.19 Standard deviation (SD):

A. SD is more difficult to calculate than quantile distribution
B. SD can have either a positive or a negative value
C. SD has the same units as the original observation
D. The variance is the square root of the SD
E. For a sample size of 15, a good estimate of population SD can be obtained by using 14 in the denominator of the equation

Q6.20 In statistical terms:

A. Events are said to be mutually exclusive when the occurrence of one, does not in any way influence the probability with which the other can occur
B. Events are said to be independent when the occurrence of one means that for all practical purposes, the other cannot occur
C. When events are independent, probabilities of one or the other occurring is the sum of the occurrence of their individual probabilities
D. When events are mutually exclusive, the probability that both will occur is equal to the product of their individual probabilities
E. The probability of an event occurring can have a maximum value of 1

Q6.21 The following statements about statistical distributions are true:

A. Poisson distribution is a type of binomial distribution
B. Binomial distribution is a type of continuous distribution
C. An F distribution is an asymmetrical distribution
D. A t distribution is a continuous distribution but with smaller tails than a normal distribution
E. A chi-squared distribution is an asymmetric distribution

Q6.22 The following statements about confidence intervals (CIs) are true:

A. CIs are a measure of hypothesis testing
B. A 95% CI in a study gives a probability of 5% of not including the estimated population parameters in that study
C. CIs help to decrease type I errors
D. A t distribution can be used to calculate CIs
E. CIs should be given for every comparison

Q6.23 The following statements about the null hypothesis are true:

A. It is a type of composite hypothesis
B. It is a type of simple hypothesis
C. It makes a statement diametrically opposite to the directional hypothesis
D. When rejected it gives the magnitude of difference between the study groups
E. It can be applied in chi-squared distributions

Q6.24 The following statements about non-parametric tests are true:

A. They are distribution free
B. They are easy to use as long as the sample size is more than 50
C. They lead to a higher probability of type I error
D. For a given sample size, they are more likely to detect a statistically significant result than a parametric test
E. A non-parametric equivalent of ANOVA is called the Kolmogorov–Smirnov test

Q6.25 The following statements about rating scales are true:

A. To imply that depression and anxiety are different points on a diagnostic scale, one would use an ordinal scale
B. Ordinal scales imply a hierarchy
C. On an interval scale, one can say that 20 feet is twice as long as 10 feet
D. No systematic relationship between different scores are implied on a categorical scale
E. A ratio scale is a type of interval scale

Q6.26 The following statements about PSE are correct:

A. It is a structured interview schedule
B. Each symptom is rated on a 5 point scale
C. It was originally designed for hospital in-patients
D. Items covering anxiety symptoms are more reliable than those covering depression
E. PSE can be used in the normal population

Q6.27 The following statements about SADS are correct:

A. It is a semi-structured interview schedule
B. There are three versions of it
C. It is mainly designed for use with hospital patients
D. It can diagnose patients with organic states
E. It cannot measure change

Q6.28 The following statements about Clinical Interview Schedules are correct:

A. It was devised by Spitzer in 1978
B. It is a partially structured interview schedule
C. It assesses symptoms in the last 1 month
D. It was designed for use with medical patients
E. It is recommend for use only by psychiatrists

Q6.29 The following statements about the GHQ are true:

A. It is a self-rated questionnaire of 60 items
B. Each question has five possible responses
C. It was designed for use in community settings
D. It can predict short-term response to various treatments
E. It has a scaled version

Q6.30 The following statements about the Brief Psychiatric Rating Scale (BPRS) are correct:

A. It contains 11 items
B. Each item is scored on a 5 point scale
C. It assesses symptoms in the last 4 weeks
D. It is unsuitable for use for patients with minor psychiatric illness
E. It is a structured interview schedule

Q6.31 The following statements about depression rating scales are true:

A. The Hamilton Rating Scale for Depression (HRSD) is a diagnostic instrument
B. The HRSD assesses symptoms in the last week
C. A score of 30 indicates severe depression on HRDS
D. The Beek Depression Inventory (BDI) is concerned exclusively with psychological symptoms of depression
E. The Montgomery and Asbeng Depression Rating Scale (MADRS) is particularly useful for assessing patients who are likely to experience marked side effects from medication

Q6.32 The following statements about the Diagnostic Interview Schedule are true:

A. It was devised by Robbins in 1981
B. It is a semi-structured interview schedule
C. It assesses symptoms that may have occurred at any time during the patient's life
D. It is possible to use it for organic disorders
E. Data collected from it are sufficient for making a diagnosis by Feighner's criteria

Q6.33 The following statements about anxiety rating scales are correct:

A. HAS was devised by Hamilton
B. HAD was devised by Snaith
C. The State–Trait and Anxiety Inventory (STAI) was devised by Spielberger
D. HAS is an adaptation of HAD
E. The Schedule for Affective Disorders and Schizophrenia–L (SADS-L) can assess anxiety

Q6.34 The following statements about mania rating scales are correct:

A. The Manic Rating Scale is rated on the basis of an 8-hour observation
B. It provides an objective measure of the severity of manic behaviour
C. It provides a measure of the frequency of manic behaviour
D. It is sensitive to change
E. It was designed for use by nurses

Q6.35 The following statements about obsessive–compulsive rating scales are correct:

A. The Leyton Inventory can produce scores for resistance
B. The Leyton Inventory can produce scores for interference
C. The Leyton Inventory can produce scores for traits
D. The Yale–Brown scale can measure only symptoms
E. The Maudsley Inventory is sensitive to change

Q6.36 The following are true of multivariate analysis:

A. It considers the relationship between a combination of two or more variables
B. Fundamentally, it involves manipulation of matrix data
C. In multiple regression, the predictor variable is being predicted from a linear combination of outcome variables
D. In linear discriminant analysis, the predictor variable has discrete levels
E. In MANOVA, the predictor variables are continuous and the outcome variables are discrete

Q6.37 The following are true of factor analysis:

A. It is used to study interrelationships among a set of variables without reference to a criterion
B. A matrix of correlations between every variable is created
C. The logarithmic combinations that best describe maximum correlation between the variables is called the principal component analysis
D. Each principal component is related to another according to a mathematical rule
E. Typically, only a few components are extracted in the initial study

Q6.38 The following are true of correlation coefficients:

A. The ϕ coefficient is used when both x and y variables are continuous
B. The ϕ coefficient is used when variable x is truly dichotomous and variable y is artificially dichotomous
C. The ϕ coefficient is used when variable x is artificially dichotomous and variable y is truly dichotomous
D. The biserial r coefficient is used when variables x and y are both continuous
E. The point biserial r coefficient is used when variables x and y are both artificially dichotomous

Q6.39 The following are true about ANOVA:

A. F distribution is the ratio of variances derived from two samples of the same population
B. It can be used to compare means of two or more samples
C. It is assumed that data from each sample should be normally distributed
D. In calculations of the F ratio, within sample variance occupies the numerator of the equation
E. The higher the value of the F ratio, the higher are the chances of rejecting the null hypothesis

Q6.40 The following methods can be used to correct confounding bias in an epidemiological study:

A. Restricting the study subjects
B. Increased matching of each case
C. Stratification of analysis
D. Multivariate techniques
E. Univariate analysis

6. Answers

A6.1
A. F There is a high degree of correlation between them
B. F The upper limits are different
C. T
D. T The range is 0.7–0.9
E. T

A6.2
A. T
B. F
C. T
D. F Girls have better linguistic ability, not necessarily vocabulary
E. T

A6.3
A. F About 25% of the total variance is explained by parental IQ
B. F Unfortunately they're not
C. F A direct relationship probably exists between IQ and socioeconomic status
D. F The reverse is true
E. T

A6.4
A. T
B. F
C. F
D. T
E. F

A6.5
A. F IQ tests can give a broad idea of brain damage
B. T
C. T
D. F The reverse is true
E. T

A6.6
A. F It captures unique configuration of traits within individuals
B. T
C. T It has a 30 item correction scale to measure how defensive a subject is in revealing his/her problems
D. T
E. F The California Psychological Inventory allows such measures

A6.7
A. F Mental ability is measured by level of problem solving and reasoning
B. F Six items are allocated to each year
C. T
D. F The Weschler Pre-school and Primary Scale of Intelligence can
E. F It allows a measurement of error of 5 points so that an IQ of 70 is considered to represent a band of 65–75

A6.8
A. F It helps in reducing socially desirable responses
B. F They also occur unconsciously
C. F The reverse is true
D. F Lie scales are included to reduce social desirability
E. F Response set measures defensiveness

A6.9
A. T
B. F It is valuable when former smokers are asked to predict the likelihood of resuming smoking
C. F It is of little value because they are strongly motivated not to be honest
D. F They are equally effective
E. T

A6.10
A. F It is an observer error
B. F They can be used
C. T
D. T
E. T

A6.11
A. T
B. F Reliability improves up to 7 points
C. F A simple analogous scale is enough
D. T
E. F Though the former is possible, the latter never is

A6.12
A. F Because the results will apply to a highly selected sample
B. F Most are, if carried out correctly
C. F It refers to assessment of any data subjectively, e.g. interpretation of EEG by simple visual examination of the traces
D. F It records qualitatively but measures quantitatively
E. F

A6.13
A. F It increases the chance of a type I error
B. T
C. T
D. F Kappa statistics are better
E. T

A6.14
A. F The reverse is true
B. F It merely becomes more acceptable than the null hypothesis until further information is available
C. F Effect size calculation only supports a hypothesis
D. F It tests the feasibility of a larger study
E. F It enables the study of the natural history of a disorder

A6.15
A. T
B. F A value of 0 implies no causation
C. F It may also be calculated in retrospective studies
D. T
E. F It implies association

A6.16
A. F
B. F It is structured
C. T By using the Index of Definition (ID)
D. F SADS can identify alcoholism and personality disorder, which PSE can't; there is no difference in reliability in identifying schizophrenia
E. F Case registers are useful for longitudinal studies

A6.17

A. F Sensitivity is an index of caseness
B. F Increasing the threshold increases specificity
C. F Increasing the threshold decreases sensitivity
D. T
E. F This may not be necessary, as a given condition may manifest itself in different ways

A6.18

A. F It is a good measure in a symmetrical distribution curve
B. F The mode is less than the median, which is less than the mean
C. T
D. T
E. F The mean is less than the median, which is less than the mode

A6.19

A. F
B. F SD is always positive
C. T
D. F Variance is SD squared
E. T

A6.20

A. F This is true for independent events
B. F This is true for mutually exclusive events
C. F This is true for mutually exclusive events
D. F This is true for independent events
E. T

A6.21

A. T
B. F Binomial distribution is a discrete distribution
C. T
D. F The tails are longer, but as the sample size increases it becomes similar to a normal distribution
E. T

A6.22

A. F It is a measure of estimation testing
B. T
C. F It helps to decrease type II errors
D. T
E. T

A6.23

A. F
B. T
C. T
D. F
E. T

A6.24

A. T
B. F The sample size must be less than 50
C. F They lead to a higher probability of type II error
D. T
E. F The Kruskal–Wallis test is a non-parametric equivalent

A6.25

A. F One would use a nominal test
B. T
C. F One can say this with a ratio scale
D. T
E. T

A6.26

A. F PSE is a semi-structured interview schedule
B. F A 3–4 point scale is used
C. T
D. F The reverse is true
E. T There is a 40 item version for use with a non-patient population

A6.27

A. F It is a structured schedule
B. T
C. T
D. F One cannot diagnose organic states using SADS
E. F It can measure change

A6.28

A. F It was devised by Goldberg in 1970
B. T
C. F CIS measures symptoms in the last week
D. F It is used in community survey
E. T

A6.29

A. T
B. F Each question has four possible responses: usual, no more than usual, rather more than usual, much more than usual
C. T
D. T
E. T

A6.30

A. F It contains 16 items: 11 by verbal report, five by observed behaviour
B. F Each item is scored on a 7 point scale
C. F There is no time limit
D. T
E. T

A6.31

A. F It is used on already diagnosed patients
B. F It assesses symptoms in the last few days
C. T
D. F MADRS is
E. T

A6.32

A. T
B. F It is highly structured
C. T
D. F
E. T

A6.33

A. F Snaith devised HAS
B. F Hamilton devised HAD
C. T
D. F HAS is an adaptation of CAS
E. T

A6.34

A. T
B. T
C. T
D. F No data are given
E. T

A6.35
A. T
B. T
C. T
D. T
E. T

A6.36
A. F Multivariate analysis considers the relationship between three or more variables
B. T
C. T
D. F Outcome variables have discrete levels; predictor variables are continuous
E. F In MANOVA, predictor variables are discrete and outcome variables are continuous

A6.37
A. T
B. T
C. F Principal component analysis uses linear combinations
D. F The principal components are independent of each other
E. T

A6.38
A. F The ϕ coefficient can be used for different types of variables, e.g. continuous, truly dichotomous and artificially dichotomous (when a continuous variable is converted into a dichotomous one) in various combinations, but not when both x and y are continuous
B. T
C. T
D. F Pearson r is used
E. F It is used when x is continuous and y is truly dichotomous

A6.39
A. T
B. F For comparing two samples, t tests are used; ANOVA is used for comparing three or more samples
C. T
D. F F ratio $= \dfrac{\text{Between sample variance}}{\text{Within sample variance}}$
E. T

A6.40
A. T
B. T
C. T
D. T
E. F

7. Social Psychology

Q7.1 The following statements about personal relationships are correct:

A. Women with more attractive partners are less neurotic
B. Men with more attractive partners are less neurotic
C. Marital satisfaction is associated with sexual satisfaction
D. There is good evidence that complementarity (dominant with submissive partner) results in a more satisfactory marriage
E. Surveys of married adults have shown that overall happiness is most strongly related to satisfaction with work, income and leisure activities

Q7.2 The following statements about gender difference in behaviour of children are correct:

A. Boys are more aggressive (physically and verbally) than girls
B. Girls are more sociable than boys
C. Differences in level of aggression in boys and girls are greater in older children
D. Girls are more suggestible than boys
E. Boys have higher self-esteem and motivation to achieve than girls

Q7.3 The following statements about prosocial behaviour are correct:

A. Prosocial behaviour is defined as behaviour beneficial to the recipient but also of some benefit to the responder
B. Helping behaviour increases when rewarded and decreases when punished
C. Helping behaviour towards a stranger is reinforcing
D. You are more likely to help someone in an emergency when you have fellow adults around than if you are with small children
E. You are more likely to take positive action while witnessing a crime if you originated from a small town than from a big city

Q7.4   The following statements are true about altruistic behaviour:

A. Individuals high in need of approval are in general more altruistic than those low in this need, irrespective of whether their behaviour is being observed
B. Irrespective of your current need, you are always likely to help others if you have a strong belief in a just world
C. In various emergency situations, females receive help more often than males
D. In various emergency situations, females offer help more often than males
E. The 'just world' hypothesis was proposed by Lerner

Q7.5   The following statements about prejudice are true:

A. Prejudice is defined as a cluster of beliefs about minority group members
B. Discrimination is hostile feelings about minority group members
C. Bending over backwards to be friendly with a minority group member is a form of discrimination
D. Decreasing contact between different racial groups can reduce prejudice between them
E. Similarities in belief are more important than racial identity when two members of different races are involved in an intimate relationship

Q7.6   The following statements about persuasive communication are correct:

A. Rapid speakers are more persuasive than slow deliberate speakers
B. The rate of persuasion is directly proportional to the physical distance between the communicator and the recipient
C. The more distracted the recipients are, the smaller are their chances of being persuaded
D. Explicit messages are more persuasive for intelligent recipients
E. There is a U-shaped relationship between the anxiety level of recipients and the fear content of a message

Q7.7   In making social judgement:

A. People tend to over-use the available base rate information
B. People tend to pay most attention to the information that supports their preconceptions
C. People estimate fairly correctly how many others would give similar judgements
D. People are usually objective about their own expectations and beliefs
E. People overestimate the role of situational factors when judging others' behaviour

Q7.8   According to Hull:

A. The interpersonal distance appropriate between close friends is 1–3 feet
B. The interpersonal distance appropriate for impersonal contact is 4–10 feet
C. The interpersonal distance appropriate for physical sport is 0–1.5 feet
D. The interpersonal distance appropriate for business contact is 4–10 feet
E. The interpersonal distance appropriate for formal contact is 2–8 feet

Q7.9   The following statements about crowding are correct:

A. Social density is defined as the number of people in a given space
B. Spatial density is defined as the amount of space available for a given number of people
C. Social density decreases as the number of people increases in a given space
D. Spatial density increases as the number of people increases in a given space
E. Increasing social density adversely affects women more than men

Q7.10   The following statements about leadership styles are correct:

A. Autocratic style is more effective in Territorial Army (TA) training
B. Democratic leaders yield greater productivity from their group when a highly original product is required
C. Irrespective of intellectual ability, a directive leader improves the performance of a group
D. In the original experiment of Lewin, groups with democratic leaders performed better than those with autocratic leaders
E. In the absence of a leader, members in a *laissez-faire* group become aggressive to each other

Q7.11   The following statements about social facilitation of behaviour are correct:

A. The presence of others enhances performance on new and exciting tasks
B. Performance on tasks in front of others improves only if it is correct
C. Incorrect performances are corrected in the presence of others
D. Performance in front of others improve because we want to look good in others' eyes
E. Performance in front of others improves because of a conflict between paying attention to others and paying attention to the task in hand

Q7.12 The following statements about group decisions are correct:

A. Common sense suggests that decisions reached by groups are more conservative than those taken individually
B. Experimental findings suggests the opposite of the above
C. Group discussion causes members to shift towards a view which is opposite to the initial view
D. One of the reasons for group polarisation is the desire of each member to outdo each other
E. Being in a group makes people become more responsible for their actions

Q7.13 The following are true about the causes of aggression:

A. According to the drive theory, aggression stems from inner tendencies
B. Blocking of ongoing behaviour is thought to be the most powerful antecedent of aggression
C. Exposure to aggressive models has higher impact on aggression than physical attack
D. People who commit extreme acts of violence have weak inhibitions against aggression
E. The psychological character of empathy is known to reduce aggression

Q7.14 The following statements about social conformity are true:

A. Asch investigated the development of social norms using the autokinetic phenomenon
B. Conformity increases in a linear relationship with group size
C. Increasing social support from another group member increases conformity
D. Festinger proposed the social comparison theory to explain conformity
E. Most of us obey authority figures because they control powerful negative sanctions

Q7.15 The following statements about attitude and behaviour are correct:

A. Measured attitudes are poor predictors of behaviour
B. There is a high correlation between the individual measures of attitudes
C. Attitudes based on personal experience predict behaviour strongly
D. Global attitudes predict behaviour strongly
E. Attitudes are evaluative; beliefs are neutral

Q7.16 The following statements about scales for measuring attitude are correct:

A. A Thurstone scale is an equally appearing interval scale
B. A Likert scale is a ratio scale
C. Semantic Differential Scales control for positional responses
D. Bogardus formulated the Social Distance Scale to measure racial prejudice
E. Moreno coined the term *sociometry* to compare the attitudes of members working in a group

Q7.17 Cognitive dissonance can be reduced by:

A. Justification of effort
B. Changing attitudes
C. Taking stimulants
D. Rationalising the information creating the dissonance
E. Counter-attitudinal advocacy

Q7.18 High dissonance is likely to occur when there is:

A. High pressure to comply
B. Low perceived choice between two actions
C. Awareness of personal responsibility for the consequences of an action
D. An expected unpleasant consequence of the behaviour for others
E. Low arousal

Q7.19 Cognitive dissonance:

A. Is motivating
B. Leads to self-doubt
C. Leads to low self-esteem
D. Is ego syntonic
E. Explains behaviour markedly at variance with the initial attitude of an individual

Q7.20 The following statements about personal space are correct:

A. Young females interact at a closer distance with males than with other females
B. People with type A personality like close interaction
C. Females are concerned with invasion of personal space from the front
D. Males are concerned with invasion of personal space from the side
E. Violent people are concerned with invasion of personal space from the back

Q7.21 The following statements about individual behaviour are correct:

A. Heider proposed the balance theory
B. If you dislike someone very much, the fact that your beliefs do not match does not alter your cognitive balance
C. If you dislike someone, you would not pay any attention to their beliefs, except when they are of the same sex as you
D. If you disagree on a particular matter with someone you dislike, you won't question your own belief
E. To correct your own cognitive imbalance, you may alter your liking of a person, although your views may not change

Q7.22 The following statements about interpersonal attraction are correct:

A. Men prefer to pair with slightly less attractive women because there is less likelihood of rejection
B. Women seek men with a similar level of attractiveness
C. Men seek women who are more attractive than themselves
D. Women seek men who are less attractive than themselves
E. Physical attractiveness has no role to play in interpersonal attraction

Q7.23 The following coping mechanisms largely correspond with the respective defence mechanisms:

A. Objectivity : rationalisation
B. Concentration only on the task at hand : isolation
C. Suppression of inappropriate feelings : denial
D. Playfulness : regression
E. Empathy : reaction formation

Q7.24 The following statements about social facilitation of behaviour are correct:

A. Performance on new tasks is facilitated in the presence of members of the opposite sex
B. On additive tasks, individuals fare better than groups
C. On conjunctive tasks, groups fare better than individuals
D. Performance on disjunctive tasks is determined by the least competent worker
E. The strength of any social influence is diluted by the number of people receiving it

Q7.25 Vulnerability to group pressure is low in:

A. Single women
B. Protestants
C. White collar workers
D. Catholics
E. The unemployed

7. Answers

A7.1
A. T
B. F There is no such association as in women
C. T
D. F People who differ psychologically do not necessarily have more satisfactory marriages
E. F Overall happiness isn't; only marital happiness is related

A7.2
A. T
B. F There is no such evidence
C. F The differences are smaller in older children
D. F There is no such evidence
E. F There is no evidence

A7.3
A. F There is no obvious benefit for the responder
B. T
C. T
D. F The opposite is true: there is diffusion of responsibility when other adults are present
E. T

A7.4
A. F They are more altruistic only if their behaviour is observed
B. F You are likely to help others only if you are currently in need
C. T
D. F Males offer more help
E. T

A7.5
A. F Prejudice is defined as a cluster of negative beliefs
B. F Discrimination is negative action against a minority
C. T This is also known as reverse discrimination
D. F Increasing contact between different racial groups reduces prejudice
E. F Unfortunately racial identity is more important

A7.6
A. T
B. F An optimum distance is required
C. F More distraction leads to a higher chance of being persuaded
D. F They work better for less intelligent recipients
E. F The relationship is an inverted U shape

A7.7
A. F People under-use base rate information
B. T
C. F They overestimate this
D. F People are not objective; this phenomenon is called illusory correlation
E. F Dispositional factors are overestimated

A7.8
A. T
B. T
C. T
D. T
E. F It is greater than 12 feet

A7.9
A. T
B. T
C. F Social density increases
D. F Spatial density decreases
E. F It adversely affects men more than women

A7.10
A. T Not always though, especially when at war
B. F *Laissez-faire* groups yield greater productivity
C. F Intellectual ability does matter
D. F They were equal in performance
E. T

A7.11
A. F Performance is enhanced on well learned tasks only
B. T
C. F Incorrect responses are more impaired
D. T
E. T

A7.12
A. T
B. T
C. F The shift is in the same direction
D. T
E. F Deindividuation occurs in a group

A7.13
A. F This is the instinct theory of aggression
B. F It's only a weak determinant
C. F There is no such comparison; both are equally important
D. F Surprisingly, they have strong inhibitions against violence
E. T

A7.14
A. F Sherif used the autokinetic phenomenon
B. F Conformity does increase but only up to a certain limit
C. F It reduces conformity
D. T
E. F Not necessarily; we obey even when they don't control such sanctions

A7.15
A. T
B. F Individual measures do not correlate highly
C. T
D. F The more specific the attitude, the more predictable the behaviour
E. T

A7.16
A. T
B. F
C. F These scales do not control for positional response
D. T
E. T

A7.17
A. T
B. T
C. F
D. T
E. T Dissonance can be reduced by forced compliance

A7.18
A. F It occurs when there is low pressure to comply
B. F It occurs when there is high perceived choice
C. T
D. T
E. F It occurs when there is high arousal

A7.19
A. T
B. F
C. F There is no such relation
D. F It is ego dystonic
E. T

A7.20
A. T
B. F They need greater personal space
C. T
D. T
E. T

A7.21
A. T
B. T
C. F Sex has no role
D. T
E. T

A7.22
A. F
B. T Both men and women seek partners of a similar level of attractiveness
C. F
D. F
E. F

A7.23
A. F The corresponding defence mechanism is isolation
B. F The corresponding defence mechanism is isolation denial
C. F The corresponding defence mechanism is isolation repression
D. T
E. F The corresponding defence mechanism is isolation projection

A7.24

A. F Sex of observers has no role
B. F Groups fare better provided social loafing doesn't occur
C. F Individuals fare better; performance by groups is determined by the least competent worker
D. F It is determined by the most competent worker
E. T

A7.25

A. F
B. F
C. F
D. F
E. F

8. Social Sciences

Q8.1　The following statements are true of the sick role:

A. It was defined by Parson as a role taken by a sick individual depending on his or her own idea of what sickness is
B. It's abnormal
C. A person given a sick role has the right to reject appropriate help
D. The sick role is a way of legitimising illness behaviour
E. It's components become invalid in drug-induced psychosis

Q8.2　The following statements are true of illness behaviour:

A. It can be described in terms of stages
B. It's culturally determined
C. Contact with a doctor is necessary to legitimise illness
D. It can be abnormal
E. It can be equated with learned helplessness

Q8.3　The following statements are true about total institutions:

A. Goffman worked in St Elizabeth's Hospital in London
B. A large ship is a type of total institution
C. In a total institution, one needs to work for items to sustain life
D. The process by which an individual becomes institutionalised is called role stripping
E. Patients in an institution formed small colonies, a process Goffman termed colonisation
F. Wing used the term institutional neurosis to describe the withdrawn state of patients in a total institution
G. Wing coined the term secondary handicap
H. Secondary handicap is seen outside an institution
I. Secondary handicap results from the unfortunate way persons with primary handicaps react to themselves

Q8.4　The following statements are about social integration are true:

A. Communes are extreme forms of normal communities
B. Durkheim published his work on anomie in 1893
C. Hooliganism in a football match is an example of anomie
D. According to Durkheim, a healthy society has a wide ranging set of values
E. Disruption of collective conscience leads to anomie

Q8.5   In Creed's study of life events (LEs) and appendicitis:

A. Patients who experienced a severe LE in the preceding year had a higher incidence of inflamed appendix than those who did not
B. Depression rates were higher in patients with an inflamed appendix
C. Patients who experienced threatening LEs had a higher incidence of non-inflamed appendices than those who did not
D. Anxiety rates were higher in patients with a non-inflamed appendix
E. There was a causal relationship between LEs and appendicitis

Q8.6   The following are types of doctor–patient relationship as suggested by Szaz:

A. Activity–passivity
B. Passivity–activity
C. Cooperation–guidance
D. Guidance–cooperation
E. Mutual participation

Q8.7   In Rosenham's experiment:

A. The researcher and his co-workers admitted themselves by reporting paranoid delusions
B. It took them an average of 3 weeks to get discharged
C. After admission, they behaved as if hallucinated
D. Those who confessed to having been ill but said they were now feeling 'better' were kept in the longest
E. Those who professed their normality were discharged earlier

Q8.8   The following statements on Durkheim's work on suicide are correct:

A. Modern studies of suicide have proved Durkheim's earlier work to be incorrect
B. Insights from Durkheim's work has been extended to other forms of illness
C. In Durkheim's work, Protestants showed a higher suicide rate than Jews
D. Durkheim argued that pre-industrial societies were characterised by high social integration
E. Durkheim argued that fatalistic suicide was produced by loose social regulation

Q8.9   Studies of sickness-absence rates have shown that:

A. They are valid measures of health and illness
B. They are affected by sick role
C. They include the unemployed
D. They measure the amount of sickness in the healthiest group of the population
E. They are limited measures of health status

Q8.10 The following are true about activities of daily living (ADL):

A. Katz formulated the ADL measures
B. ADL measures the major everyday functions only
C. Having a bath is measured as a major function
D. The responses are added to give an overall score of abilities
E. Doing up zips and buttons is considered to be a major activity

Q8.11 Studies of illness behaviour have shown that:

A. Rates of medical consultation accurately reflect the frequency of symptoms experienced
B. Older people with aches and pain consult their doctors more frequently than younger ones
C. Normalisation of some symptoms blocks the emergence of important diagnostic information
D. Most patients go to doctors immediately after they perceive a symptom
E. Very few of the symptoms perceived are self-limiting
F. Seeking medical help is almost always related to severity of illness
G. Patients with known heart disease go to doctors quicker than those without when they experience chest pain

Q8.12 In traditional culture, sick roles are:

A. Universal
B. Diffuse
C. Ascriptive
D. Particular
E. Achieved

Q8.13 The following help to break down the sick role:

A. An increasing incidence of chronic conditions
B. Medicalisation of social problems
C. Ageing of the population
D. Self-help movements
E. Renewed interest in preventive medicine

Q8.14   The following statements are true about the medical profession:

A. It enjoys greater autonomy than other professions
B. It is more likely to be a terminal occupation
C. It is relatively free of lay evaluation
D. Student doctors go through more liberal socialisation than other students
E. Most legislation concerned with the profession is shaped by legal authorities

Q8.15   Morbidity studies have shown that:

A. Married men have higher rates of strokes than single men
B. Single men have lower rates of heart disease than married men
C. Married women are more at risk of developing mental illness than single women
D. Despite changes in the social role of women, the overall morbidity figures remain the same
E. Married men have higher rates of lung cancer than single men

Q8.16   According to the OPCS data (1994):

A. The average household size in the UK is 2.45
B. Lone-parent families make up 10% of all families
C. One-person families make up 10% of all families
D. Typical families make up 40% of all families
E. Twenty-three per cent of families are couples with no children

Q8.17   The following are risk factors for marital breakdown:

A. Childlessness
B. Marriage for ten or more years
C. Both partners being younger than 20 years at the time of marriage
D. The bride being pregnant at the time of marriage
E. Having five or more children under the age of 11

Q8.18   The following statements about social class are true:

A. Social class is the same as social status
B. Higher class people are superior to lower class people
C. Lower class people believe in a social hierarchy
D. Middle class people believe in social mobility
E. In a crisis, middle class people stick together more readily than lower class people
F. Middle class people sympathise with those needing social assistance more readily than lower class people

Q8.19 The following statements are correct:

A. Working class people consult health services more frequently than middle class people
B. Working class people have higher availability of health resources than middle class people
C. Working class people have greater health needs than middle class people
D. Working class people consult their doctors more often for sickness certificates than middle class people
E. Availability of good medical care varies directly with the health needs of the population

Q8.20 Surveys have shown that:

A. Middle class patients have shorter consultations with their doctors than working class patients
B. Families of unskilled workers are more likely to have strong role segregation between husband and wife
C. Families of unskilled workers are more likely to stereotype children in terms of gender
D. The risk of death after retirement is twice as high in social class 5 as in social class1
E. Rates of longstanding illness are twice as high in unskilled workers as in professionals

Q8.21 According to Rack, *Gastarbeiter*s are:

A. People from developing countries who migrate to developed counties
B. Middle-aged men
C. Migrating men with no intentions of returning to their countries of origin
D. Migrating families seeking better opportunities
E. Rural people leaving the countryside for economic reasons

Q8.22 Jarmen indices:

A. Are measures of social deprivation
B. Were devised by the sociologist Norman Jarmen
C. Are devised from data derived from epidemiological surveys
D. Include the elderly living alone
E. Include mobility

Q8.23   The following are true about illness and social causation:

A. Japanese men who emigrate to the USA are more likely to die of stroke than those who stay in Japan, where ischaemic heart disease (IHD) causes more deaths
B. American men are more likely to die from stroke than from IHD
C. People in social classes 1 and 2 have a higher mortality rate from IHD than those in classes 4 and 5
D. Compared to others in the UK, there is a higher rate of IHD among Asians in the UK
E. Compared to others in the UK, there is a higher rate of hypertension and stroke among Afro-Caribbeans in the UK

Q8.24   The following are true about life events and mental illness:

A. Studies involving the Schedule of Recent Events (SRE) have shown that as many as one-third of events are missed
B. The fall off rate of the Life-Events and Difficulties Schedule (LEDS) is around 5%
C. The SRE uses the concept of life change units
D. The LEDS uses a 4 point scale of threat
E. The SRE scoring system assumes that the effects of life events are additive

Q8.25   The following are correct:

A. The relationship between increase in the number of life events and relapse of schizophrenia holds true only for severe events
B. The relationship between increase in the number of life events and relapse of depression holds true for all types of event
C. The brought-forward time for depression is about 9 months
D. The brought-forward time for schizophrenia is about 3 weeks
E. There is a consensus that severely threatening life events are associated with the onset of organic disorders only when there is concurrent psychiatric disorder

8. Answers

A8.1
A. **F** The sick role is given by the society
B. **F**
C. **F**
D. **F** Doctors legitimise illness
E. **T**

A8.2
A. **T**
B. **T**
C. **F**
D. **T**
E. **T**

A8.3
A. **F** St Elizabeth's is in Washington
B. **T**
C. **F** Items are automatically provided
D. **F** It is called the mortification process
E. **F** Colonisation was a process whereby the patients pretended to show acceptance of institutionalisation
F. **F** Barton coined the term institutional neurosis
G. **T**
H. **T**
I. **F** Secondary handicap results from others' reactions

A8.4
A. **F**
B. **T**
C. **F**
D. **F** Durkheim felt that a healthy society has only a few common sets of values
E. **T**

A8.5
A. **F** These patients were more likely to have non-inflamed appendices
B. **F** Patients with a non-inflamed appendix had higher depression rates
C. **F** They had a higher incidence of pathologically inflamed appendix
D. **F** This was not studied
E. **F** Only an association was found

A8.6

A. **T**
B. **F** This was suggested by Friedson
C. **F**
D. **T**
E. **T**

A8.7

A. **F** They reported hallucination the night before
B. **F** They were discharged, on average, after 30 days
C. **F** They behaved completely normally
D. **F** They were discharged earlier
E. **F** They were kept longer

A8.8

A. **F**
B. **T** For example, the influence of social integration on health
C. **T** Because of lower social integration
D. **T**
E. **F** Fatalistic suicide was caused by over-regulation

A8.9

A. **F**
B. **F** They are affected by illness behaviour
C. **F**
D. **T**
E. **T**

A8.10

A. **T**
B. **F** it also measures minor functions
C. **F** This is classed as a minor function
D. **T**
E. **T**

A8.11

A. **F** Several other factors are considered before going to doctors
B. **F** Because aches and pains are considered normal
C. **T**
D. **F**
E. **F** Most of them are
F. **F**
G. **F**

A8.12

A. **F** They are particular
B. **T**
C. **T**
D. **T**
E. **F** They are ascribed

A8.13

A. **T**
B. **T**
C. **T**
D. **T**
E. **T**

A8.14

A. **F**
B. **T** Once a doctor, a person is likely to continue to be one
C. **T**
D. **F** Their socialisation is more stringent
E. **F** It is shaped by the profession

A8.15

A. **F** Married men have lower rates of stroke
B. **F** Single men have higher rates
C. **T**
D. **T**
E. **F** Married men have lower rates

A8.16

A. **T**
B. **T**
C. **T**
D. **T**
E. **T**

A8.17

A. **T**
B. **T**
C. **T**
D. **T**
E. **F**

A8.18

A. F Class is determined more by life chances, attitudes and values, whereas status is determined more by occupation, skills, other behaviour, etc.
B. F Only patterns of behaviour differ
C. F They believe that the society is divided into 'us' and 'them'
D. T They believe that mobility depends on utilisation of individual abilities
E. F That's a lower class perspective
F. F They do not; lower class people do

A8.19

A. F Working people consult less frequently
B. F
C. T
D. T
E. F It varies inversely

A8.20

A. F Their consultations are longer
B. T
C. T
D. T
E. T

A8.21

A. F
B. F They are young men
C. F These are called settlers
D. F
E. T

A8.22

A. T
B. F Brian German, a GP
C. F They are devised from census data
D. T
E. T

A8.23

A. F

B. F More Japanese men in Japan die from stroke and more American men in America die from IHD, but more Japanese men in America die from IHD

C. F People from lower classes have lower mortality rates from IHD

D. T

E. T

A8.24

A. T

B. F It is 1%; it is 5% for SRE

C. T

D. T

E. T

A8.25

A. F The relationship holds for any type of life event

B. F The relationship holds only for severe events

C. F It's about 1–2 years

D. F It's about 10 weeks

E. T

Bibliography and References

1. Caplan, H. I. and Saddock, B. J. (Eds) (1995). *Comprehensive Textbook of Psychiatry*, Vol. I, 6th edn, Williams & Wilkins, Baltimore.
2. Snell, R. S. (1987). *Clinical Neuroanatomy for Medical Students*, 2nd edn, Little Brown, Boston.
3. Morgan, G. and Butler, S. (Eds) (1993). *Seminars in Basic Neurosciences*, Gaskell, London.
4. Trimble, M. E. (1996). *Biological Psychiatry*, 2nd edn, Wiley, Chichester.
5. Lader, M. and Herrinton, R. (1990). *Biological Treatments in Psychiatry,* OUP, Oxford.
6. Puri, B. K. and Tyrer, P. J. (1992). *Sciences Basic to Psychiatry*, Churchill Livingstone, Edinburgh.
7. Weller, M. and Eyesenck, M. (Eds) (1992). *The Scientific Basis of Psychiatry*, W. B. Saunders, London.
8. Freeman, C. and Tyrer, P. (Eds) (1992). *Research Methods in Psychiatry: A Beginner's Guide*, 2nd edn, Gaskell, London.
9. Henderson, A. S. (1990). *An Introduction to Social Psychiatry*, OUP, Oxford.
10. Armstrong, D. (1994). *Outlines of Sociology as Applied to Medicine*, 4th edn, Butterworth-Heinemann, Oxford.
11. Tantum, D. and Birchwood, M. (Eds) (1994). *Seminars in Psychology and Social Sciences*, Gaskell, London.
12. Byron, R. A. and Byrne, D. (1977). *Social Psychology: An Understanding of Human Interactions*, 2nd edn, Allyn and Bacon, Boston, Massachusetts.
13. Atkinson, R. I., Atkinson, R. C., *et al.* (1993). *Introduction to Psychology*, 11th edn, Harcourt Brace, London.
14. Kendell, R. E. and Zealley, A. K. (1994). *Companion to Psychiatric Studies*, 5th edn, Churchill Livingstone, Edinburgh.
15. *British National Formulary*, No. 33, March 1997.

The Author

Since nursing for four years at the London Hospital, Belinda Pratten has been involved in Community Health issues as a researcher/writer and activist. She is currently Assistant Secretary at Bloomsbury CHC, a member and ex-Chair of Health Rights Management Committee and co-editor of *Health Matters* magazine. Belinda lives in Tower Hamlets, where she is Chair of the local Health Campaign.

'Physical and social deprivation directly affects women as carers, workers and patients. All women have the right to good health and access to quality health care, regardless of age, race, disability or sexual orientation. We also have the right to a healthy environment including safe, secure, warm, dry homes and communities free from violence. Equal access to healthcare must be based on respect, being listened to and understood. Through the Women's Health Charter we seek to overcome some of the present obstacles and lay claim to our health rights.'

from: Tower Hamlets Women's Health Charter

Contents

ACKNOWLEDGEMENTS

I would like to thank all my colleagues at Health Rights and particularly Beverley Beech, Karen George and Jane Cowl for their support and constructive comments; Myra Garrett at Tower Hamlets Health Campaign for lending me the Campaign 'archives'; Wendy Savage for commenting on an earlier draft; Liz Sheppard for her moral support at the end of the 'phone and Ann Treneman, whose practical assistance was far beyond the call of friendship. Finally my special thanks to Alex and Casey Farquharson, whose weekly visits provided a welcome distraction from the rigours of writing.

MEDICAL GLOSSARY

ANALGESICS
- pain-killing drugs

EPISIOTOMY
- a cut made by the midwife or doctor to prevent tearing of the skin during birth

HAEMORRHAGE
- extensive bleeding

INTRAUTERINE
- within the womb (uterus)

IN VITRO FERTILIZATION (IVF)
- artificial fertilization of the female egg by male sperm outside a woman's body, in a laboratory – the 'test tube baby'

MORBIDITY
- illness/disease

MORTALITY
- death

NEONATAL
- relating to the first year of life

OBSTETRICIAN
- a doctor who specializes in childbirth

OBSTETRICS
- the medical management of childbirth

PAEDIATRICIAN
- a doctor who specializes in the medical care of children

PATHOLOGICAL
- relating to, involving, or caused by disease

PERINATAL
- relating to the first week of life

PERINATAL MORTALITY RATE (PMR)
- the proportion of babies who die in labour or immediately after birth

PULMONARY EMBOLISM
- a blood clot affecting the supply of blood (and oxygen) to the lungs

ULTRASOUND
- a scan of the womb to show the fetus using sound waves

FOREWORD

The 'Wendy Savage case' as it was usually called, occupied much media and medical attention for many months, as the circumstances leading to the suspension of an established woman obstetrician by her professional male colleagues were unravelled, examined and pronounced upon in a long drawn out public enquiry. The outcome of the enquiry – exoneration of the charge of professional incompetence – was followed by Wendy Savage's own account of the case[1], which highlighted some of the wider issues involved. On the whole, however, the case was treated by public, press and the medical profession as concerning questions of *individual* behaviour and attitudes in the particular local circumstances of maternity care in Tower Hamlets, albeit located within the highly politicized context of the management of childbirth in the technologically complex culture of the 1980s. Given such a context, in which warring factions repetitively contend both the essential normality and abnormality of childbirth, such clashes of judgement and personality are bound to happen – or so the argument went.

As this book shows, however, the 'Wendy Savage case' depended for its formulation and momentum on far more than local personality conflicts. At least four issues of fundamental importance to the social handling of childbirth were involved. First, and perhaps most obviously, there is the question of what childbirth is – how it is defined and seen, whether there is conflict or consensus in the views of different social groups. Is childbirth part of life or a task of medicine, a component of medical or of maternal labours? Must the way childbirth *is* managed derive its authenticity from the experience of mothers, or from the needs of medical professionals to justify their own claim to expertise?

The anthropologist Margaret Mead pointed out many years ago that the reproductive division of labour between the sexes can never simply be considered 'natural', for every society defines what is 'natural' in its own way.[2] In modern industrial societies the 'naturalness' of childbirth came to include medical management and control sometime in the early years of the twentieth century.[3] Childbirth ceased to be largely women's business, conducted at home and drawing on the expertise of experienced women in the community, and became instead something that women took to doctors and to hospitals. Perhaps most important in this process was the promotion of an uncritical association between *safety* and *medical authority*. Women came to understand that a safe delivery was one supervised by obstetricians.

The second issue highlighted by the Wendy Savage case is thus the one emphasized in her own account of the circumstances surrounding the enquiry; that of *control*. When childbirth is subject to a medical definition, it is doctors who must control childbirth not women. The implication of this – in a world where women are anything other than merely accessories to the male professional point of view – is inevitably to some extent a struggle for power. At times quiescent and at other times openly confrontational, this dispute about control manifests itself in many different ways. There may be concern about territory; home or hospital? Attention may be drawn to patterns of staff-client communication, and within this to the medical tendency to demote women from their autonomous status as human beings to a mechanical one as containers of uteruses. Psychological aspects of childbirth are commonly stressed as of lasting but ignored importance, so that for example, the need for mothers and babies to become emotionally attached to one another is seen to pre-empt other more narrowly physiological considerations. More recently the challenge of user-organisations in the childbirth field has concentrated on a critique of the increasingly interventionist nature of modern obstetrics, a feature of particular significance in the Wendy Savage case.

The third and possibly most fundamental issue embedded in Wendy Savage's suspension is that of *science*. If Wendy Savage's obstetrics was less interventionist than that of her male colleagues, was it also less scientific? At times this seemed to be the claim that was being made. Scientific medicine was equated with interventionist obstetrics in a linear model which assumes that the invention and routine use of technology is what characterizes human evolution in general. This central claim of modern obstetrics rightly receives a good deal of attention in this book, which argues that the *representation* of obstetric medicine as a science is considerably more powerful and has contributed much more to its success, than its *actual* scientific basis would or should merit. Much that is routinely done or contended in modern obstetrics still has the status of fashion or heresay; very little is based on the rigorous controlled evaluation of different approaches or strategies.[4]

Finally, and in a way most problematic of all, there is the question of the gendered division of labour itself. The people who give birth are always women and the obstetricians are usually men. Wendy Savage is one of that rare breed, a woman-identified professional. Most professional high-achieving women absorb the ideologies of their male peers and lose any identification they may have had with women as a group. The fact that Wendy Savage retains and articulates a belief that women themselves should be in control of childbirth does not make her popular with most male obstetricians. From this point of view, the ideological stances adopted in the Wendy Savage enquiry are reminiscent of a much earlier struggle between female and male midwives for ascendancy over childbirth,[5] and indeed with arguments about the relative social positions and rights to supremacy of men and women.[6] Not coincidentally, the arguments put for and against Wendy Savage's (relatively) non-interventionist obstetrics are also themes in that other crisis of modern obstetrics: the role and work of midwives.[7]

In short, Wendy Savage's suspension, exoneration and continuing struggle to practise humane, client-sensitive obstetrics, is about much more than the local circumstances of obstetrics in Tower Hamlets. For this reason, this book is to be welcomed as exposing a number of basic and enduring issues of maternity care to public scrutiny.

Ann Oakley
1989

REFERENCES

(1) W. Savage, *A Savage Enquiry*, London, Virago, 1986.

(2) M. Mead, *Male and Female*, Harmondsworth, Penguin, 1962.

(3) A. Oakley, *The Captured Womb*, Oxford, Blackwells, 1984.

(4) I. Chalmers, M. Enkin, M. Kierse (eds), *Effective Care in Pregnancy and Childbirth*, Oxford, Oxford University Press, 1989.

(5) J. Donnison, *Midwives and Medical Men*, London, Heinemann, 1977.

(6) S. Rowbotham, *Women, Resistance and Revolution*, London, Allen Lane, 1962.

(7) R. Devries, *Regulating Birth*, Philadelphia, Temple University Press, 1985.

INTRODUCTION

'In learned conferences doctors may share their doubts with each other, but these rarely reach the outside world and so we accept what medicine offers us on its terms. The message is that medicine can cure us. The myth is so powerful that we tend to overlook the evidence – not just from research but from our own lives – that its achievements are very much more modest.'[1]

The suspension of consultant obstetrician Wendy Savage for alleged professional incompetence and her subsequent exoneration following a public enquiry, saw some of these doubts reach the outside world. It was probably the first such hearing to be held in public and this was important, for it allowed the contending medical and professional arguments to be aired openly before a wider audience, namely the users of health care services. As Mrs Savage herself stated, it showed women that choices exist regarding the type of care they receive.

For Mrs Savage it was important that the enquiry proceedings should be held in public. Her reputation was at stake, and as Michael O'Donnell reported in the *British Medical Journal*, she had already undergone 'trial by gossip' in medical circles, involving both professional and personal smears.[2] Even if a private enquiry had completely exonerated her, questions would have remained. As it was, the enquiry was not merely held in public, but in a blaze of publicity.

In the media, the story appeared to be a straightforward confrontation between 'technology' and 'nature', centring around a personal conflict in the Department of Obstetrics and Gynaecology at the London Hospital. In fact, the issues are far more complex: fundamentally the story is about power, about the rightness and applicability of medical science to define

and control women's reproductive capacity and hence womanhood itself.
For Ann Oakley, the medicalization of childbirth has been a process
whereby:

> '...a particular area of social behaviour (pregnancy) comes to be
> separated off from social behaviour in general and reconstituted as a
> specialist, technical subject under the external jurisdiction of some
> expert authority.'[3]

Childbirth (as with fertility control in general) has been taken out of its
social context, where it exists as an integral part of women's lives, and
transformed into an isolated medical event, to be defined and treated
within the context of the disease-oriented framework of western scientific
medicine. For this to happen, pregnancy and birth had to be viewed as
primarily a hazardous event, one that necessitated medical and technological intervention and treatment. Only by focusing on its potential for pathology could it be viewed as a legitimate area of medical concern. Importantly, the medical profession has had the full backing of state policy with
regard to this. This is not to imply that the aim of reducing maternal and
perinatal mortality and morbidity (death and illness of mothers and babies
in childbirth) is not praiseworthy. Indeed, throughout most of this century
this has been a concern of women's health movements which have cam-
paigned for more hospital services to be made available to women.[4] However, as these services have developed and the move towards 100 per cent
hospital care has become a reality, it has also become evident that these
services have often failed to take into account women's needs. Women are
expected to fit into the medical model rather than vice versa, and this has
seriously undermined their ability to take charge of this crucial aspect of
their lives.

This is not to deny the important contribution of medical intervention to
the reduction of mortality and morbidity during pregnancy and childbirth.
Yet there is a major difference between treating complications which may
arise during pregnancy and defining the entire process in terms of pathology. Birth is not an illness, it is essentially a natural, healthy, biological
process which is part of a woman's life experiences, as Ann Oakley and

Hilary Graham describe it:

> 'It is akin to other biological processes (like menstruation) that occur in a woman's life. This is not to say that it is a woman's "natural destiny" to bear children. Rather for those who do, the process is rooted in their bodies and their lives and not in a medical textbook.'[5]

When birth is viewed as being akin to illness it is essential that women are exposed to an all-risk monitoring process by those who are experts in illness. When it is viewed as being a part of women's lives, the expert's role is reduced, the obstetrician must therefore play a supportive, rather than controlling role. This is the basis of the 'woman-centred' approach to obstetrics, an approach which is adhered to by Wendy Savage, who believes in as little intervention as possible, consistent with safety and who has worked to take antenatal care to women through community-based rather than hospital-based services. Yet it was argued that her reluctance to intervene in this natural process exposed patients to unnecessary risk and resulted in the allegations of incompetence which led to her suspension.

The issues raised by the 'Savage affair' therefore centre around who defines, and thus controls, pregnancy and childbirth. It questions the role of medicine in relation to the wider society – and reveals its limitations. It has shown how medical science does not consist of indisputable facts, but that the type of services provided depend on which facts are used, how they are applied and in whose interests; it has revealed the political nature of decisions concerning health and particularly the sexual politics of health care. Finally, it has shown what little influence users have either over their own care or over the decision-making process regarding the provision of services as a whole. These are the issues which are addressed in this book.

REFERENCES
(1) J. Mitchell, 1984, p.10
(2) *BMJ*, May, 1986
(3) A. Oakley, 1984, p.1
(4) A. Oakley, 1984
(5) A. Oakley and H. Graham, 1981, p.53

1

THE WENDY SAVAGE AFFAIR

On 24 April 1985 Wendy Savage was suspended from her post as Honorary Consultant in Obstetrics and Gynaecology to Tower Hamlets District Health Authority (DHA), under the Department of Health procedure HM (61) 112, accused of professional incompetence. This procedure is rarely used – and had never before been used in Tower Hamlets – but is designed to protect patients should a clinician be considered unsafe to practise. It is designed to be used in cases of mental or physical ill health, or alcohol or drug dependence. Under the terms of this procedure a doctor is not allowed to treat patients pending an enquiry into the allegations made against her or him. This may have implications for the clinical teaching of medical students, and on this pretext Mrs Savage was also suspended from her duties as Senior Lecturer to the London Hospital Medical College. Her suspension marked the culmination of a history of long standing disagreements between herself and her consultant colleagues, Gedis Grudzinskas, Trevor Beedham and John Hartgill.

That these disagreements were not taken into consideration before it was decided to suspend Mrs Savage is one of the many question marks that remain over this affair. The decision to hold an enquiry was taken solely by the Chair of Tower Hamlets DHA, Francis Cumberlege, on the advice of Gordon Bourne, then a consultant obstetrician at St Bartholomew's Hospital and Regional Assessor for maternal mortality. Even so, under the terms of HM (61) 112 Cumberlege was not then obliged to also suspend Mrs Savage from her teaching duties. That was a decision which was to prove costly to the authority and which so angered local women and GPs that they quickly rallied to Mrs Savage's support.

15

The Case for the Prosecution

The charges against her questioned her competence with reference to five cases where her treatment was described as having been 'bizarre', 'idiosyncratic' and 'unorthodox'. It was argued that taken together these five cases showed:

- a 'consistent aberration of clinical judgement', which exposed these patients to 'unnecessary and unjustified risks'.

- a failure to acknowledge the consequences of such judgements and hence to modify practice accordingly.

- an unwillingness to recognise that there was cause for concern in her response to these criticisms and therefore 'a readiness to avoid and shift personal consultant responsibility for such shortcomings as she is prepared to acknowledge existed'.

- that her approach risked confusing junior staff, thereby undermining and jeopardising consistent standards for safety and continuity of care within the Department.

Thus it 'was and is not such as meets the standards reasonably required of a senior lecturer at a major teaching hospital'.

(Taken from: Terms of Reference to the Enquiry, statement of case.)

It was implied that Mrs Savage put her own strongly held attitudes and beliefs before the health and welfare of her patients. Four of the five cases required emergency Caesarian sections, three of which involved breech presentations, one having had a previous Caesarian delivery, another, twins and the third had a trial of labour. The other cephalic presentation (ie. head first) was stillborn. The core of the allegations was that she delegated too much responsibility to the General Practitioners (GPs) for antenatal care and that she should have intervened at an earlier stage during labour. During the enquiry these cases were referred to by their initials, which I have continued to do here where relevant.

In July 1985, Mrs Savage took the case to the High Court for breach of contract, claiming that she had been unjustly suspended and demanding immediate reinstatement. In September, the Court ruled that a professional could best be judged by her peers, a judge could not presume to comment on a complicated issue such as obstetrics and therefore the case should be heard before an enquiry set up by the Health Authority. (This is interesting, given that judges often 'presume' to make pronouncements on other equally specialist subjects.) However, the ruling stated that the enquiry should take place within 'a reasonable period of time'.

The Media's Response

As soon as they heard of Mrs Savage's suspension, local GPs took steps to ensure that their outrage at this turn of events, and that of her patients and students, was made public. *The Guardian* was the first to take up the story – and indeed was strongly supportive throughout, particularly during the ordeal of the public enquiry. On 7th May 1985, two weeks after her official suspension, it reported a protest organised by these GPs and involving midwives and many local women in support of Mrs Savage. By June the story had caught on, not least as a result of her supporters' efforts to ensure justice was done, and *The Observer* was already suggesting that her case had become a national *cause célèbre*.

The media's response was largely sympathetic to the obvious injustice of Mrs Savage's case. Yet at times it seemed that the way in which the story was taken up so widely owed less to a sudden conversion to the cause of women's rights, or even to that of issues in childbirth, than to the way it fitted in to the soap opera style so loved by the British press. It had all the elements of a good news story: a hospital drama involving a lone doctor supported by her patients against the weight of the establishment. That the doctor involved – and her patients – were women and the establishment dominated by men inevitably gave rise to that other much loved media scenario, the Battle of the Sexes. On the one hand there was an outspoken, divorced mother of four, socialist and feminist, with strongly held views on women's rights in childbirth; on the other, her more conservative male medical colleagues (with allegations of that other all-male fraternity, the Freemasons, thrown in). A typical portrayal is that of Richard Lindley in *The Listener* (6.2.86):

'Wendy Savage is different from her consultant colleagues in almost every way. Where they are interested in doing private work in Harley Street, she is only concerned with the National Health Service. Where the Professor wants to get the hospital involved in the new and glamorous world of test-tube babies, she seems more interested in running an abortion clinic. Where the other consultants summon their antenatal patients to the hospital, she goes out to the community...While senior medical men dress soberly, she wears apricot dungarees and a T-shirt. And where they are men, she is a woman.'

Although there was largely sympathetic coverage of the issues involved, the response to Mrs Savage herself was rather more ambiguous and the portrayal of her personality often obscured these issues. *The Observer* described her as 'an abrasive campaigner' and every article had some such similar epithet. She was 'forthright', 'outspoken' and 'determined', and with the media's fondness for leaders she became the 'Childbirth Movement' personified. Perhaps only as a leader could a strong, if not forthright, woman be portrayed with any sympathy. It was also noted how unlike a doctor she was in her approach to patients, in her dress – and in her gender. But such things were merely commented on, leaving the reader to judge their merits, yet it is unlikely that these comments would be passed on a man in her position. This ambiguity was perhaps due to the fact that Mrs Savage does not fit into the usual media stereotypes of women. In the same way, Mrs Savage's politics were stated, as was her 100 per cent support for the NHS, but in none of the national daily or Sunday newspapers were the issues raised as being inherently political. Only magazines such as *Spare Rib* and *New Society* took up the political dimension as opposed to the personal; indeed, in April 1986 *New Socialist* described the enquiry as 'the most significant political trial of this century'.

Philosophical Issues in Obstetrics

The enquiry took place in February 1986. Both sides brought in expert opinion to advise on the way these five cases had been handled. In his opening speech, Ian Kennedy, QC for Tower Hamlets DHA, stressed that the enquiry was not concerned with wider issues of medical and obstetric philosophy:

> 'The question is not about the principle. The question is about how it has been put into practice in these five cases, and it is an enquiry not about theories but about dangers in obstetrics. Of the five cases, one resulted in a neonatal death and another in a stillbirth...In the remaining cases, the management was outside all normally accepted principles...'

However, as the enquiry was to reveal, it is not possible to discuss the practice without understanding the principles behind it. In all areas of medicine there exists some degree of disagreement over aspects of clinical management, but this is particularly the case in obstetrics, where pregnancy and childbirth are as much a psycho-sexual experience as a purely biological event. One of the most contested issues within this debate is the appropriateness of elective (ie. planned) surgical delivery (Caesarian). There is a broad spectrum of opinion on this issue: at one end it is believed that all women who have had one Caesarian section should automatically be advised to have one on a subsequent pregnancy, although in this country it is usually considered necessary only when a woman has had two previous Caesarians. Opinion is also divided as to whether a woman whose fetus is lying upside down, (ie. a breech presentation) should always require elective (planned) surgery, or whether she should be offered a 'trial of labour'. This means that she is allowed to attempt to give birth vaginally and an emergency (unplanned) Caesarian performed only if this proves impossible. This view sees the wishes of the woman herself as being paramount, providing that this is consistent with the safety of both mother and child – and has been endorsed by the World Health Organisation.

A further point of contention is the length of time labour should be allowed to continue without surgical intervention. At one end of the spectrum it is suggested that labour should be no longer than 12 hours, but many obstetricians disagree with this. Instead it is argued that since not all labours progress at an 'average' rate and as it is not always possible to determine the exact time of onset of labour, the length of time does not matter providing that neither the mother or fetus are showing signs of distress. More important is the fact that the woman feels that she is in control of the process and is allowed the possibility of giving birth in the way in which she would prefer. However, it would appear that as the safety of intervention procedures improve, there is a marked reluctance for doctors to do nothing, in case something goes wrong. In this way defensive intervention may become departmental policy:

> 'There is a subtle influence in obstetrics to absolve a doctor who intervenes in the course of a normal pregnancy and that, by implication, exposes his conservative colleague to censure for inactivity when an infant is born dead. This places a premium on intervention as a form of insurance for the doctor, although the consequences are detrimental for some patients.'[1]

The Findings of the Enquiry

The existence of a broad spectrum of obstetric philosophy was acknowledged by the enquiry when the panel reported its findings in July 1986. In the Report of the Enquiry they stated that an act, or failure to act, could only be described as incompetent if it failed to come 'within the broad limits of accepted medical practice'. The panel then went on to say that:

> '...the test is not whether a particular witness, or the medical members of the panel, would themselves have done, or not done, a particular thing in a particular way. If there is a reputable school of medical thought which would, or might, act in a certain way then that way will

be within the broad limits of accepted medical practice.'[2]

The report accepted that her colleagues were concerned with Mrs Savage's practice, but argued that even a genuine concern was not necessarily a justified concern and stressed the importance of professional autonomy. All consultants are equal and each has a right to practise in the way they think best. Therefore one could not be criticized for favouring one school of thought over another. While recognising that at a teaching hospital it is important that students and junior staff should be clear about the basis on which judgements are made, they argued that this necessitated extra responsibilities for those at senior level and not departmental conformity, which would not be in the interests of academic freedom.

Overall, the enquiry panel maintained that: 'The characteristics of Mrs Savage's practice are care and consideration for the patient'.[3] Although criticisms were made regarding particular aspects of management of four of the five cases, the overwhelming majority of charges were declared completely invalid. The panel stated that the decision to share care with the GP was not unreasonable and in fact had no bearing on subsequent events. Only in one instance (AU) was management considered to be near the bounds of accepted medical practice, but while the panel agreed that it had been 'idiosyncratic', it was not incompetent. They accepted that the trial of labour had been the expressed wish of both parents, and the evidence of paediatricians, called in to testify as to the likely cause of death in the absence of a post-mortem, stated that this baby's death was not connected with the way in which either the pregnancy or the delivery was managed. It is regrettable that the probable cause of death of baby U was not established until the enquiry, almost two years later. This long delay and their involvement in the proceedings against Mrs Savage must have convinced the parents that she was to blame and can hardly have helped to alleviate their grief over the loss of their child.

The Enquiry Report also referred to suggestions that had been made outside of the enquiry that Mrs Savage was a victim of some form of conspiracy. In reply the authors of the Report stated that evidence was not brought forward to refute or confirm these allegations, which was not part of their terms of reference. However, they expressed a concern that there was an 'undercurrent of opposition' from some of her colleagues that may have meant that they were 'unconsciously unable to give her the benefit of the doubt'. Given that this was not a factor put forward in her defence it is significant that it should be referred to in the report. Therefore it is necessary to look at the background to these proceedings and in particular at how a case could be established without Mrs Savage being aware that any investigation was underway.

A History of Conflict

According to Ian Kennedy, QC, the need for some kind of investigation into Mrs Savage's practice came to light because of concern expressed by junior staff and her consultant colleagues. It would appear that the subsequent proceedings were instigated by Gedis Grudzinskas and Trevor Beedham. According to evidence given at the enquiry, Dr Richards, the District Medical Officer, wrote to Professor Grudzinskas in August 1984, stating:

> 'We need a report on the competence of the consultant concerned to carry on in clinical obstetric care. The initial reports we had from Mr Beedham indicated that there was considerable doubt on this matter. The authority needs to be assured that the clinician is safe to continue managing obstetric patients.'[4]

Neither the Professor nor Mr Beedham were willing to give such assurance. The case of AU was sent to Gordon Bourne; in the following nine months, the notes for the other four cases were also sent to him for assessment. That these cases could be considered representative is by no means evident, as Mrs Savage has argued:

> 'Approximately 3 per cent of women will have a breech presentation at term, one in a hundred has twins, and only one woman in a thousand has twin breeches.

22

In the Mile End branch of the London Hospital, the Caesarian section rate...was about 10 per cent in 1983-4, although in 1984 my personal rate was just over 8 per cent. So clearly these cases had been carefully chosen – 80 per cent of this small group of women had been delivered surgically and 60 per cent had breech presentations!'[5]

Both in evidence to the enquiry and in the media, paediatric senior lecturer and consultant Peter Dunn stated that NO obstetrician could be judged on the basis of their five worst cases. Similarly, Michael Joffe argued in the *British Medical Journal* that such a small, hand-picked group inevitably distorts a practitioner's overall record. He suggested that although the basis of these charges was insufficient intervention, there have been instances where intervention itself has 'led to disaster'. Therefore the consequence of either policy cannot be judged by only a few instances, a point also made by Iain Chalmers, Head of the National Perinatal Epidemiology Unit, in *The Lancet*:

'In making comparisons in this way it goes without saying that inferences about the relative merits and demerits of particular policies should not be based on individual cases selected to make a point but on representative and sufficiently large samples of the cases managed by the consultants being compared.'

Both commentators suggested that if her colleagues were primarily concerned with the issue of safety, a systematic evaluation of clinical practice would have been more appropriate than disciplinary proceedings.

Moreover, that Bourne should have been considered an impartial assessor is also in some doubt. Not only do St. Bartholomew's and the London hospitals share a joint Academic Department of Obstetrics and Gynaecology, which meant that he worked closely with both the Professor and Mrs Savage, but his own views on obstetric practice clearly favour the orthodox, interventionist school, so it was far from likely that he would be sympathetic to Mrs Savage's approach.

Once Bourne had established that there was a case to answer, Francis Cumberlege, chair of the DHA, took the decision to suspend Mrs Savage without informing either herself or the full DHA. Mrs Savage had been unaware that any investigation was taking place. Although Grudzinskas had repeatedly asked her for case reports, which she gave him on the advice of the Medical Defence Union, he had never directly questioned her competence nor asked for her opinion as to why she had conducted these cases as she had. Much later an internal enquiry into the affair by Tower Hamlets DHA suggested this decision had been ill-conceived:

> 'It appears that the long standing, and well known disagreements and differences within the Obstetrics Department were not taken into account in this matter.'[6]

Such disagreements were revealed in evidence to the enquiry. Professor Taylor, honorary consultant obstetrician at St Thomas' Hospital (and an opponent of Mrs Savage on the issue of abortion), stated that he had met Grudzinskas at a social gathering soon after his appointment to the London Hospital in 1983, when Grudzinskas remarked that one of his first tasks was to change his senior lecturer (ie. Mrs Savage). Although Grudzinskas stated that he could not remember making such a remark, the enquiry accepted that he had and that it had caused Taylor sufficient concern for him to remember. The panel stated that although his feelings at this time (1983) were not relevant to the five cases in question, they may have had some bearing on later events.

In early 1984, Grudzinskas wrote to Mrs Savage criticizing her for failing to coordinate her activities with the rest of the department, after she had spoken out against proposed changes to maternity service provision. He also stated that her first loyalty should be to the interests of the Department, the Academic Unit and hence to him personally as the head of that Unit. Both in this letter and in a second, sent a month later, he warned her that if she continued to show dissent she may be dismissed. He was also critical of her community work on the grounds that it took too much time from her teaching responsibilities.[7]

It seems evident that the debate between different policies at the London Hospital turned into open conflict, perhaps fuelled by personality clashes. What is less evident, however, is the way in which an essentially internal dispute over medical policy could proceed to a full-scale disciplinary hearing without any attempt having been made to resolve it internally. After the first major difference with the Professor in 1984, Mrs Savage suggested that they referred the matter to the 'Three Wise Men', a panel of senior consultants whose role is to arbitrate and attempt to resolve such conflicts. Grudzinskas rejected this as being inappropriate; it would appear that he felt it far more appropriate to go for her complete dismissal.

Although 'conspiracy' cannot be proven, there is at least some suggestion that Mrs Savage was not exactly a welcome member of the Department and that the question of her dismissal had been raised before the initial complaint. Yet a full-scale enquiry is both costly and damaging to all concerned and in this case its findings were such as to cast doubt as to whether it should have been used at all. Certainly this was the view of the local community whose public support for Mrs Savage was more than a match for the hospital establishment. It is probable that no-one expected it to result in a full-scale enquiry. It is perhaps significant that Grudzinskas was a lecturer at St Bartholomew's Hospital in 1979 when the HM (61) 112 procedure was used against Pauline Bousquet, whose tragic story has only recently come to public attention.

Returning to Work

Despite her total exoneration in July 1986, Mrs Savage was unable to return to work until the following November. Obviously the divisions within the Department before her suspension had deepened considerably during this time and the other consultants expressed their unwillingness to continue working with her, to the extent of threatening to resign. Therefore the DHA was forced to set up a second panel, with Dame Alison Munro as Chair, to recommend a framework in which to re-establish professional working

25

arrangements. This panel worked on the basis that Mrs Savage had been reinstated in her honorary NHS consultant contract and that any question of her clinical competence had been disposed of. However, as Dame Alison revealed in a letter to the Chair of Tower Hamlets DHA:

'The four consultants practising in the Department since Mrs Savage's suspension say the Department is working well without Mrs Savage and will not accept our two starting points which we have regarded as the inescapable basis of our work. They have therefore been unwilling to discuss the detail of the proposed arrangements. This is despite close and personal discussions with each of them not only by the panel but also by the Chairman of the Medical Council, the outgoing and incoming Dean of the Medical College and senior colleagues from other disciplines.' (6.10.86)

The authority accepted the panel's recommendations and Mrs Savage's colleagues were forced to work with her again; none of them resigned as they had threatened to do. However, Mrs Savage has not been able to return to work in the Department of Obstetrics and Gynaecology. She was based in the Department of Clinical Epidemiology, where she had continued to do research work during her suspension, until November 1987, when she transferred to the Joint Academic Department of General Practice and Primary Care. It is clear that since her return her colleagues have not been exactly welcoming; according to one staff nurse, if Mrs Savage enters the ward office when they are there, they get up and leave. Apparently they even complained when she was invited to the ward Christmas party! Mrs Savage has won the battle to clear her name but remains the focus of her medical colleagues' animosity.

After the enquiry, Tower Hamlets DHA set up its own internal review of this affair. Its findings were reported to a closed session of the DHA meeting in June 1987. Although these findings have not been made public, a copy of the report was sent to the *Independent*. Thus it was revealed that the report was highly critical, stating that the enquiry should never have taken place. The complaint from Mr U should have been properly investigated under the *complaints* procedure. Instead Grudzinskas proposed to hold an enquiry, Mrs Savage was excluded from the subsequent investigations and the case inevitably became a matter for *disciplinary* procedures.

Even so, these procedures were not correctly followed: after investigation into the first case (AU) the Chairman took the view that there was no case against Mrs Savage, and that should have been the end of the matter. Moreover, she should have been warned that a disciplinary enquiry was under consideration and given the opportunity to answer the allegations.

This internal report was rejected by the Authority, 8 votes to 7, after the Chairman used his casting vote, on 11 June 1987. Yet it would seem that even at this stage all was not above board. At the May meeting of the DHA it was proposed to hold a vote on this report, but, against previous practice, the Chair refused to accept a verbal motion and insisted it was put in writing. At the next (June) meeting he had filled two vacant places on the Authority (which had been vacant for months). Both new members voted with the Chair despite their lack of knowledge of the case, as did the consultant nominee to the Authority, who morally should have declared an interest since he had been heavily involved in the proceedings. Local authority members of the DHA – who had in the past been vocal in their support of Mrs Savage – were unable to attend, since it was election day. Therefore the decision to delay the vote drastically affected the outcome. Even so the result was close, and that it should have rested on the casting vote of the Chair is highly questionable given his own instrumental role in the affair. As a result DHA member Shelbourne, one of the authors of the report, resigned on the grounds that he could not be effective as a member of the DHA if he was unable to challenge the status quo.

To many, perhaps, this bewildering affair is past history, an unfortunate incident best forgotten, particularly by those within the London Hospital who put in motion this chain of events. Yet it is evident that the internal rumblings of dissent rumble on within the Department of Obstetrics and Gynaecology – and may soon come to a head with the centralization of all

27

maternity care at Whitechapel and the closure of the Mile End unit. For Mrs Savage the fight continues, but the fight is not hers alone. It involves all women, and those who care for women in pregnancy and childbirth, who question the prevailing status quo when it comes to the issue of who controls childbirth.

REFERENCES
(1) O'Driscoll et al, *BMJ*, 1975, no.4, pp.272-9
(2) Report of the Enquiry, 1986, p.11
(3) Report of the Enquiry, 1986, p.13
(4) W. Savage, 1986, p.46
(5) W. Savage, 1986, p.85
(6) *Independent*, 9.7.87
(7) W. Savage, 1986, pp.31-35

DEVELOPING COMMUNITY SERVICES

Tower Hamlets has been described as the most deprived health district in England and Wales.[1] The borough has the highest level of unemployment in Inner London (18.8 per cent compared with 12.8 per cent for Inner London as a whole), with only 24 per cent of the population in non-manual occupations, compared with a national average of 43 per cent. Poverty, poor housing, overcrowding and homelessness are problems endemic to the area and, as with unemployment, local ethnic minority groups are worst affected. There are a number of ethnic minorities in Tower Hamlets, including those of Somali, Afro-Caribbean, Chinese and Vietnamese origin and 20 per cent of the population are Bengali.

There has been a wealth of evidence associating high levels of mortality and morbidity with social deprivation; indeed, Tower Hamlets was made special reference to in the 1980 Black Report, *Inequalities in Health*. This is no less the case in relation to pregnancy and childbirth, where there are persistent disparities between the mortality rates for different social classes. Thus the background of many local women would automatically label them high risk, on statistical evidence alone. Mrs Savage's insistence that risk is an individual factor and not a statistical one, gives them a choice that they may not otherwise have.

Together with the active support of Peter Huntingford, Grudzinskas' predecessor, Wendy Savage has played an important role in improving and developing reproductive care services to meet the needs of the community. In 1977, Huntingford and Savage started a day care abortion service, to avoid the wastefulness of unnecessary hospital admission, making the procedure less traumatic. With this they developed an efficient and caring counselling service, using non-medical counsellors and providing them with full support. They also encouraged GPs to become more involved in the care of pregnant women, providing support and facilities for both 'Domino' delivery (where the woman is delivered in hospital either by the GP or the community midwife and leaves six hours after birth) and for home births. In 1982 Mrs Savage furthered her commitment to shared care by holding antenatal clinics in three local health centres; in effect, she went into the community rather than expecting the community to come to her, making antenatal care more accessible, more acceptable and more flexible to its recipients. Moreover, studies have shown that most women want the option of seeing a woman doctor.

Although it became apparent that these moves were unpopular with her consultant colleagues, they undoubtedly were popular with those it involved - GPs, midwives, and the women themselves. Within a week, and before it was officially announced that Mrs Savage had been suspended, there were strong local protests. 149 medical students (75 per cent of those canvassed) signed a petition, paying tribute to her conscientiousness as a teacher and her commitment to her patients. Community midwives demanded her reinstatement, praising her high standards of care, the good relationships she built with both staff and patients and the valuable support she gave to midwives, stating that her support of women's choices was inextricably linked to the promotion of good midwifery care in Tower Hamlets. There was also strong support from local GPs – 68 of the 84 GPs in the district (85 per cent) signed a letter expressing their concern as to the detrimental effects Mrs Savage's suspension would have on the provision of maternity services and demanding her reinstatement. As Dr Edmondson, one of the organisers of this protest, stated in *The Guardian*:

> 'This letter speaks volumes on the way in which Mrs Savage is regarded in this district, particularly since alleged malpractice is involved, which is a very serious complaint. Despite this, the doctors have had enough confidence in her to put their names to this letter.'

Yet possibly more significant than the response of these professionals was the active and vocal support of the women of Tower Hamlets. Women from all sections of the community were willing to speak out and attend marches and demonstrations. The first march, to the DHA meeting on 13 June 1985, was held six days into the Ramadan fast, but this did not deter large numbers of Bengali women from taking part. One patient, Rita Hotson, was widely quoted in the media as stating: 'If I could get on top of Nelson's Column with a megaphone and shout Dr Savage's praises, I gladly would.'

Three of her most ardent supporters were women whose management during pregnancy and delivery was being used to support the allegations of incompetence. Linda Ganderson, whose first baby had been stillborn, Denise Lewis and Susan Payne were upset and angry that their 'cases' had been used in this way. They were happy with the treatment they had received and it had been important for them to be involved in taking decisions and being empowered to make informed choices for themselves. Both Linda and Susan pointed out that they had chosen Wendy to care for them in their subsequent pregnancies and Denise stated that she would do so, should she have another child.

Importantly, a predominantly working class area with a large ethnic population such as Tower Hamlets is perhaps not the natural constituency for such protest. The issue of choice in childbirth has primarily been taken up by articulate, middle-class women, possibly more aware of the lack of power over the conditions in which they give birth because they have some degree of control and choice over the conditions of their lives. But silence does not mean assent, as Ann Oakley argues:

'The assumption of the satisfactory meaning of silence is to be found not only in obstetrics but in many other fields; indeed, it is generally characteristic of the ideological stance of dominant groups towards oppressed minorities.'[2]

By involving local women and enabling them to take an active role in their pregnancies and births, Wendy Savage offered women a choice which they did not want to have taken away. As Christine Smith, then Chair of the Community Health Council, explained in a letter to the DHA:

'It has been most heartening to see a consultant with a willingness to identify how local women will best use a service and then attempt to implement it to the benefit of patients, staff and local GPs.'

The Case for Community Obstetrics

Sheila Kitzinger, the doyenne of the alternative birth movement, was quoted in *GP Magazine* (25-5-85) as saying that Wendy's suspension was a direct attack on the concept of community obstetric care, echoing the concern of many Tower Hamlet GPs. Without consultant support the task of community obstetrics may be virtually impossible. Since 1982 Wendy herself has provided consultant cover for general practitioner obstetricians. This model of care was not new but followed the pioneering work of Dr Kenneth Boddy in the Sighthill district of Edinburgh and Dr Luke Zander in the London Borough of Lambeth. It was found that antenatal care based on general practice improved the continuity of care that women received and increased patient satisfaction. There is also some evidence that it may reduce perinatal mortality.[3]

Orthodox medicine insists that a woman attends the hospital for routine check-ups throughout her pregnancy – she must make a 'booking' visit soon after her pregnancy has been confirmed by her GP, then must attend every month to 28 weeks, every fortnight to 36 weeks, then weekly to term. The availability of sophisticated diagnostic techniques makes it essential for this type of care to be delivered in the hospital context. This approach stresses the importance of these visits for the health of the baby, as Bourne asserts, despite the lack of good evidence:

...the earlier a pregnant woman presents herself to a professional advisor for her first antenatal visit the higher is the chance that she will have a perfectly normal pregnancy and baby. Also, the longer she defers her first visit, the greater are her chances of having some complication or a dead baby.'[4]

Obviously antenatal care is important as a means of checking out whether a woman is at potential risk but this is a classic case of 'shroud-waving', ie. if you don't do what the doctor says, your baby might die. The baby is not going to die simply because you did not attend the hospital. In 1985 the World Health Organisation stated categorically that there is no evidence to suggest that routine antenatal care has a positive effect on the outcome of pregnancy. Similarly, in 1985 GP Graham Marsh argued in the *British Medical Journal* that:

'Traditional, routine antenatal care...is no longer relevant to the female population of the 1980s and its continuation is a meaningless and unnecessary waste of resources. The string of normal findings painstakingly recorded on cooperation cards in hospitals and general practice must suggest that some consultations must have been a waste of time for the midwife, doctor, and especially the mother.'[5]

In 1980, a study of maternity services in Aberdeen, where they were thought to be exemplary, found that this type of care was a poor predictor of risk and that, in fact, more could be learned from listening to women, taking a comprehensive medical history, than from technological diagnostic methods. For example, 56 per cent of abnormally small births had not been detected antenatally, while for every case correctly diagnosed over twice as many were ascribed to a healthy fetus. At the same time, 456 women were diagnosed as being hypertensive, subsequently this was found to be wrong in 256 cases. Such incorrect diagnoses undoubtedly cause much stress as well as subjecting women to unnecessary investigations and admissions, when in fact most justified admissions were emergencies which could not have been predicted. The authors also suggest that too much emphasis was placed on the value of technology: in 80 cases a correct diagnosis was made for 67 women on the basis of a history alone, in only six cases did this change after a physical examination and in only seven, after technological investigation.[6]

These findings certainly do not justify the imposition of a treatment oriented model of care on all pregnant women as routine.

Dr Marsh argues that the number of consultations could be dramatically reduced without decreasing predictive accuracy. In his own practice he found that when each attendance is carefully structured and women are given more time, the result is a more individual and cost-effective style of care. Huntingford also suggests that the present number of visits are largely unnecessary; a woman need only be seen four times during the course of her pregnancy to ensure all is well, unless abnormalities arise or she herself feels the need for more. He directly challenges the assumptions on which routinized hospital care is based:

> 'The cost-benefit for this process is the death of fewer babies and the survival of more healthy ones. But the evidence that the dehumanizing rituals of antenatal care designed to ensure compliance should take the credit is doubtful.'[7]

These 'dehumanizing rituals' are often at best irrelevant, at worst oppressive for the women involved; reinforcing their own feelings of inadequacy as they wait for hours for a brief contact with perhaps a different doctor offering contradictory advice each time. Moreover, doctors have a tendency to treat women as 'irresponsible children', incapable of taking decisions concerning their own welfare or of understanding the procedures they have to go through and thus are offered little explanation.

Blaming the Patient

Yet where take up of antenatal care is low this is often seen in terms of ignorance on the part of the mother or even as representing an ambivalent attitude towards her pregnancy. Ann Oakley has shown how this has always been the response of health care professionals and policy makers to this problem and how the small impact of antenatal care in the 1920s and 1930s was also perceived in this way.[8] The belief in the dominance of medical science and in medicine's own claims for itself automatically assumes that the problem must lie with women. Therefore the Government's response has been to focus on health education campaigns aimed at pregnant women to encourage them to attend, but as Helen Roberts argues:

'There is unlikely to be a substantial body of women in the UK who do not know of the existence of antenatal clinics. There is a substantial body of women who do know of their existence but who are unwilling or unable to travel long distances over town by public transport with younger children in tow, who cannot afford the long waits, who do not like sitting around with their knickers off, or who, since they have never had the purpose of their visits and their tests explained to them, believe there is none.'[9]

In 1979, research by Hilary Graham and Lorna McKee found that more than half the women they interviewed were unhappy with the care they received at the antenatal clinic. They suggest that the attitudes of professionals were often unhelpful and uncaring and that the clinics tended to be geared towards the needs of doctors rather than those of the mother-to-be. Further evidence of this was found in a similar study by Ann Oakley in 1981. For both Oakley and Graham there is a fundamental difference between the medical and maternal perspectives of antenatal care.[10] They found that doctors tended to be highly dismissive of the very real fears and anxieties of women unless they appeared relevant to their own search for pathology. Serious requests for information tended to be treated casually by doctors, instead women were given reassuring platitudes or insufficient explanations which often led to more anxiety and confusion. Many women said they felt unable to ask questions. In one group, (at the London Hospital) women asked on average one question per encounter; in another (in York) 40 per cent said they did not feel they could ask questions, and of those who said they could in theory, many did not in practice. It was evident that doctors defined the limits of these encounters on the assumption that they knew best, for example:

'Doctor: (Reading case notes) Ah, I see you've got a boy and a girl.

Patient: No, two girls.

Doctor: Really, are you sure? I thought it said...(checks in notes) oh no, you're quite right, two girls.'[11]

If a woman cannot be relied upon to know the sex of her two children, it is obvious that she could not be trusted to take an active part in her own care.

Similar results were obtained from a survey of antenatal patients by Tower Hamlets Health Campaign in 1985. This asked women who had experienced both full hospital antenatal care and shared care between the hospital and their GP how they felt about each type of care. These women overwhelmingly favoured the care they received at the health centre. Specifically they identified the different appointment systems as a major factor. At the Whitechapel Hospital a block booking system operates, resulting in long waits sometimes of up to five hours and when it was their turn to be seen they felt rushed, that they should leave quickly because the next patient was waiting. They also stated that the hospital was like a 'conveyor belt' or a 'cattle-market' where there was no continuity of care and often further long waits for different procedures. Again it was felt that doctors at the hospital clinics did not encourage women to ask questions. As one respondent stated, they were there to tell you what to do, not to answer your questions or discuss anything. If women stated their anxieties they tended to be dismissed with a 'not to worry' or 'it's not important' with no further explanations or 'respect for their intelligence'; as another said, at the hospital she 'got too many oohs and ahhs and not enough explanations of what was going to happen'.

At the health centre the women did not have to make a long journey, nor wait for hours as individual appointments were made, which also meant that that time was for that woman alone, so she felt less hurried. The surroundings themselves were more familiar and they usually saw the same doctor or midwife at each visit. This was perceived as being extremely positive as it allowed the women to feel more confident about asking questions and to feel that the professional had a better understanding of them and their situation (Tower Hamlets Health Campaign – unpublished).

These studies suggest that hospital-based care is perceived as an alienating experience for many women and therefore non-attendance or delayed booking may be as much a judgement on the type of care on offer as a reflection of the level of awareness of its importance.

Graham and McKee argue that social factors are an important determination of risk, but that those mothers who may be defined as most at risk may be the ones who find it most difficult to attend, whether because of lack of transport, childcare responsibilities, or language barriers. Evidence from the National Perinatal Epidemiology Unit suggests that women attending late or sporadically for antenatal care tend to be younger, or to have had many children already; more often single; from social classes four or five or unemployed, or are immigrants; factors which statistically place them at risk. Therefore in an area such as Tower Hamlets with a high birth rate, poverty, poor housing and a large ethnic population of child-bearing age, it is particularly important that appropriate antenatal care is provided by improving the scope and quality of community services, rather than attempting to 'educate' women into coming into the hospital. The aim must be to make these services acceptable and accessible to pregnant women.

Community Obstetrics in Tower Hamlets

In 1985, a number of local voluntary organisations, in association with Queen Mary College, sponsored an intensive enquiry into the standard of health in Tower Hamlets and the appropriateness of health care services in this district. In evidence to this enquiry women's health groups and GPs expressed their concern at the effects of Wendy's suspension on community services. As one GP stated:

'We are most concerned about the suspension of Mrs Savage ... she is deeply committed to improving safety and women's experiences through a commitment to community care and where appropriate,

37

non-intervention in childbirth. She is also our only female specialist in an area where all the customers are female ...

Our patients are extremely concerned about her suspension ... this includes a number who for cultural reasons do not want to see a man, but many more.'

Another GP gave evidence of the results of a clinical audit into the effectiveness of community-based care that was undertaken in 1983. This showed that for the 390 GP/community bookings in Tower Hamlets between 1974 and 1983 the perinatal mortality rate was 7.8 per 1,000 total births; the rate for the whole of Tower Hamlets had been 14.5 per 1,000 in 1982. The Caesarean section rate for women booked by GPs was 5.1 per cent, half that for women booked in hospital (10.4 per cent) and forceps were required four times less frequently (excluding breech deliveries). Although this does not indicate that community booking by itself reduces mortality or the need for intervention, it does suggest that effective selection procedures had been used to identify women who were unlikely to be at risk and for whom routine intervention was unnecessary. It also found that community booking significantly reduced delay in attendance for antenatal care.

In 1982-3, a survey of 115 women who delivered while registered with South Poplar Health Centre showed that 44 per cent of hospital booked women booked at 16 weeks or more gestation and 28 per cent at 20 or more weeks. By instituting a policy of booking women at GP surgeries, in which Mrs Savage was the supervising consultant who attended the surgeries in person, very considerable reductions were made in the proportion of women who booked late. Only 11 per cent of women booked at the surgery were at least 16 weeks pregnant at booking and only 6 per cent were 20 or more weeks: a statistically significant reduction.[12]

These results show that the development of community obstetric services has made an important impact on the take-up of antenatal care. Possibly the most fundamental difference between hospital and community based services is the assumption on which each is founded. In the former, care is given within a theoretical framework that is concerned with the treatment of disease, so that decisions are likely to be oriented towards intervention; whereas the latter is geared towards the promotion of health and therefore to prevention. For women the distinction is crucial; by taking maternity care out of the treatment-oriented model it takes pregnancy out of the realm of sickness, where it is subject to medical definition and control, and moves it nearer to becoming seen as part of women's healthy lives.

REFERENCES
(1) eg. M. Joffe, 1985; B. Jarman, 1986
(2) A. Oakley, 1984, p.243
(3) eg. I. Chalmers, 1984, *Journal of the Royal Society of Medicine*, no.77, pp.340-2
(4) G. Bourne, 1984, p.123
(5) *BMJ*, 1985, vol.291, p.646
(6) Hall et al, *The Lancet*, 12-7-80, p.78
(7) J. Rakusen and N. Davidson, 1982, p.7 (foreword by P. Huntingford)
(8) A. Oakley, 1984
(9) H. Roberts, 1981, p.10
(10) A. Oakley and H. Graham, 1981
(11) A. Oakley and H. Graham, 1981, p.66
(12) J. Robson, K. Boomla and W. Savage, 1986, *Journal of the Royal College of General Practitioners*

3

THE ISSUE OF SAFETY

'The debate is often seen as a conflict between "technology" and "nature" but in reality I think it is more complex than that. All women want the best possible chance for their babies and will endure crowded, hot antenatal clinics, physically uncomfortable monitoring in labour and operative delivery if the doctor advises that this is best for the baby.

What I think women, most midwives and some doctors are increasingly questioning is the scientific evidence on which today's obstetricians base this advice and secondly, the application of a mechanistic type of care to all women as routine.'[1]

The case against Mrs Savage was based on the issue of safety. Her colleagues were apparently concerned that her practice was not only 'unorthodox', but also unsafe. In this chapter I wish to look more closely at this issue in relation to orthodox obstetrical practice and in particular whether this practice actually lives up to its own claims for itself.

The history of western obstetrical care has been characterized by the use of increasingly advanced technological and pharmacological monitoring and intervention throughout pregnancy and birth. It is believed that this has significantly contributed to a reduction in maternal and perinatal mortality and morbidity. However, these rates have tended to be the only criteria used to assess the value of such advances. There has been far less concern in medical literature to evaluate the *quality* of maternity care, to assess maternal satisfaction with the services provided. Relatively little attention has been paid to the considerable indirect evidence that suggests the major reason for the continuing fall in both rates may be social, rather than medical intervention.

Because of this, medicine has tended to develop a system of obstetric care which requires much conformity to ensure that all women receive essential advice, tests and supervision. The assumption that *all* women are at risk of developing complications during their pregnancy necessitates such a system; the issue of safety is therefore central to the medicalization of birth. Yet inevitably it fails to take into account women's individual needs and ignores the subjective experience of becoming a mother. It has been suggested that since the risks of childbirth have been dramatically reduced over the last 50 years and because it is now possible to detect and treat complications in their early stages, 'risk' has become only one of a number of criteria to be assessed in the evaluation of care – and patient satisfaction with its quality is as relevant as rates of mortality and morbidity.

Medical science has defined pregnancy and birth as an abnormal state requiring intervention and treatment and hence as an appropriate and legitimate area for medical concern. In this way normal physiological occurrences are defined as 'symptoms' and the entire process is seen as being potentially hazardous, which is then reflected in the way in which society views it. Doctors have the power of definition: by defining the problem they can control the solution. When it is presumed that no birth is normal except in retrospect it follows that all births should benefit from the available technology, which is the perceived solution to the problem of pathology.

The Hospitalization of Childbirth

The medicalization of birth is associated with its hospitalization, aligning normal childbirth with the care of the sick and placing it firmly under medical control. The establishment of lying-in hospitals for pregnant women first occurred in the seventeenth century, but for many years their contribution was such that in 1871 Florence Nightingale was to remark that:

'...There is a large amount of preventable mortality in midwifery practice, and that, as a general rule, the mortality is far far greater in lying-in hospitals than among women lying-in at home.'[2]

Without the knowledge or techniques to further their role, hospital confinements at first offered little of therapeutic value and greater risk of infection. However, it did enable doctors to learn about childbirth and the pursuit of knowledge concerning the fetal condition and the physiological processes involved in pregnancy and childbirth has been integral to the obstetricians' claim to expertise. Therefore techniques to further this knowledge have been actively sought.[3] Possibly the first major development was the invention of surgical forceps towards the end of the nineteenth century, which marked the beginning of the era of interventionist treatment. Over the last 50 years the development of sophisticated technologies and advances in biochemical understanding, such as hormone function and genetics, have made pregnancy amenable to the mechanistic model of medicine and facilitated medical control over the birthing process. With these developments doctors no longer needed to rely on an intermediary, ie. the pregnant woman, and could thereby claim superior knowledge and expertise. Having defined birth in this way, the search for pathology becomes the purpose of all maternity care, so that the subjective feelings of the pregnant woman are less important than the objective measurement of the physiology of pregnancy and the fetal condition.

Questioning the Scientific Evidence

The development of maternity services in Britain, as in other western industrialized countries, has evolved around this assumption of pathology and the belief that routine monitoring and medical management have significantly contributed to the decline in maternal and perinatal mortality rates. It has been a crucial factor in the development of state policy in this area. The findings of a series of government reports have endorsed the move towards a centralization of maternity services in large, obstetric hospitals with 100 per cent hospital confinement. For example, in 1970 the Peel Committee stated:

'We believe that the greater safety of hospital confinement for mother and child justifies the objective of providing sufficient hospital facilities for every woman.'[4]

In the 1920s approximately 15 per cent of births took place in hospital; by 1958 this number had increased to 60 per cent; by 1981, 94.2 per cent of deliveries occurred there.[5] That there has simultaneously been a fall in perinatal mortality rates has reinforced the belief that there is a causal relationship between the two. This was the conclusion of the Maternity Services Advisory Committee Report in 1984:

'The practice of delivering nearly all babies in hospital has contributed to the dramatic reduction in stillbirths and neonatal deaths and to the avoidance of many child handicaps.'

Yet the available statistical data suggests that this relationship is likely to be as much coincidental as causal. The evidence to suggest that place of birth has been a particularly significant factor in the reduction of mortality is not as strong as may at first be supposed. In 1976 the *British Medical Journal* reported on a survey of births in Cardiff: despite the fact that the number of home births decreased from one in five in 1965, to one in a hundred in 1973, there was no corresponding statistically significant decrease in perinatal mortality, nor in the number of possibly preventable deaths.[6]

A study by Fryer and Ashford, reported in the *British Journal of Social and Preventive Medicine* in 1975, showed similar results.[7] They analysed the effect of hospital confinement on perinatal mortality for each successive year between 1956 and 1973. Between 1956 and 1967 there was a positive correlation between local authorities with above-average hospital confinements and below-average Perinatal Mortality Rates (PMR), although lessening each year. However, further rises in hospital births from 1968 did not sustain these results; indeed, local authorities with above-average hospital rates had above-average mortality. While this in no way indicates a causal relationship between these variables, it would suggest that the routine application of hospital care to greater numbers of women does not mean greater benefit.

43

Probably the most extensive analysis of such data has been undertaken by Marjorie Tew, a research statistician whose initial results so antagonised professional interests that she was subjected to personal ridicule and forced to do much of the work using her own time and money. In both 1958 and 1970 the National Birthday Trust carried out two national perinatal surveys. Although extensive data from these surveys were officially published, this did not include comparative analysis of place of birth and mortality, although this had been included in the survey. The assumption that reproductive technologies lessen the risk to both mother and child (and therefore that the safest place to give birth is in a specialized obstetric hospital with the necessary equipment to hand) would mean that hospital mortality rates would be significantly lower. Tew found that, in fact, the exact opposite was true: using the National Birthday Trust survey's own definition of risk (the antenatal prediction score), at every level of risk the specific PMR for each group was higher by far for those born in consultant obstetric units than those under the care of a GP or midwife. The perinatal mortality rate for consultant units was 28 deaths for every 1,000 births, compared with only 5.4 per 1,000 born under the care of the GP or midwife. Even when corrected for age, number of previous children and social class, a statistically significant difference remained.

For Tew, modern maternity care is based on a theory that has never been justified in practice. She believes that the data suggests that the active obstetrical management of birth may actually increase the risk to some infants. She argues that the central issue is not safety, but professional self-interest and suggests that the fall in mortality rates over the last 30 years has been primarily due to the improved general standard of health of the population:

'There is abundant evidence to show that women today are healthier, to the extent that they have withstood the dangers of obstetric management.'[8]

Researchers Rona Campbell and Alison MacFarlane have agreed with

Tew's findings, but argue that her conclusion again suggests a causal relationship which has not been proven. Their own comprehensive evaluation of the available data led them to conclude that 'there is no evidence to support the claim that the safest policy is for all women to give birth in hospital'. They also found some inconclusive evidence to suggest that morbidity may be higher among mothers and babies cared for in an institutional setting.[9]

Evaluating the Technology

Iatrogenic illness is that which arises as a result of medical treatment, such as the side effects of certain drugs. The difficulty of proving the existence or otherwise of iatrogenic illness resulting from the use of monitoring procedures and invasive technologies used in pregnancy and childbirth, rests in the fact that these methods have tended to be introduced routinely without any prior systematic evaluation either of their effectiveness or of potential costs. Peter Huntingford has stated that:

'New methods are introduced and routinely applied as rapidly as changing fashions in the clothes we wear. Research and evaluation of the new technology in maternity care is often poor; the results conflicting and doubtful; continued examination of obstetric routines is desultory and made difficult because there are few attempts to record information and results, especially with regard to the long-term future.'[10]

Once technologies have been introduced and the demand for their use has been created, there is a marked reluctance to undertake a cost-benefit analysis in the form of a randomized, controlled trial, on the basis that this would mean depriving women of procedures which are currently available. But while technological innovation has been unhampered by the brake of clinical evaluation, the march forward has not been without its setbacks.

The tragedy of thalidomide in the 1970s is well known: a drug widely supplied to help women to sleep during pregnancy was found to have caused limb deformities in the unborn fetus. Results such as this have led doctors to be much more cautious in prescribing drugs for pregnant women, although analgesics and anaesthetics continue to be given (often routinely) to women in labour. Yet while there are certain legal restrictions placed on the introduction of new drugs, requiring rigorous testing to minimise the risk of such harmful effects, there are no similar restrictions or requirements for the introduction of new technologies. For example, ultrasound scans are now used routinely and a woman may be scanned several times during her pregnancy. These scans provide an important aid in detecting twins or potential problems such as intrauterine growth retardation or malformation. It is clear that there are no short-term dangers involved in their use, but there has never been any attempt to evaluate possible long-term effects. Therefore safety cannot be proven and there have been some studies to suggest that such long-term effects may exist,[11] and it would seem that there is no benefit to be gained from its routine use. In fact both the WHO and the Department of Health recommend against using it routinely.

There is no doubt that technology can be beneficial, if not life-saving, but without proper evaluation no procedure can be said to be risk free. Thus their routine use in the absence of clinical indicators may in fact increase risk for some women.

However, given that the detection and prevention of complications is the major priority of all maternity care, it is necessary for these procedures to be used routinely, 'just in case'. The acceptance of technological intervention as standard practice has meant that even in retrospect normal birth has been redefined. The use of electronic fetal monitoring, induction or acceleration, or epidural anaesthesia will all be entered on the notes as 'normal delivery'. Indeed, the medical profession has such faith in this that in 1974 the Medical Defence Union booklet *Consent to Treatment* stated:

'The Union does not consider that a maternity patient need give her written consent to any operative or manipulative procedures that are normally associated with childbirth. When she enters hospital for her confinement it can be assumed that she assents to any necessary procedure, including the administration of a local, general or other anaesthetic.'

Given that most women are denied the option of giving birth anywhere other than in a hospital, they are immediately caught in a Catch-22 on the basis of this assumption. Although this statement was subsequently withdrawn after protests from the Association for Improvements in the Maternity Services (AIMS), it is evident that the underlying attitudes remain. The message is clear, 'doctor knows best'. Doctors have received specialist training making them expert on all matters pertaining to pregnancy and childbirth. By not heeding their advice, the woman is not risking simply her own health, but that of her baby; therefore as Bourne advises pregnant women:

'You will learn to adapt to the difficulties and to accept the changes that occur. You will learn to cooperate with your professional advisers...'[12]

Intervention and the Experience of Childbirth

The physiology of childbirth is such that it cannot be reduced to the functioning or otherwise of parts of a machine, but is a complex process, or rather a number of processes, so perfectly attuned to each other that intervention may upset this balance. Psycho-social needs, such as the need for a supportive emotional environment, contribute to this balance and are as important as clinical needs. The medicalization of birth has often resulted in its mechanization, whereby the woman is perceived and treated as a reproductive machine to which other machines may be attached, so that, as Sheila Kitzinger argues:

'Some doctors treat each pregnant woman as if she was an ambulant pelvis and every woman in labour as if she were a contracting uterus.

47

There are many who assume that they know what is best for a woman better than she can know herself.'[13]

Similarly, Barbara Rothman has argued that the separation of mind and body which characterizes the medical viewpoint has been extended to the woman and her fetus, who are then viewed as being in competition rather than as an integral unit. This implies that the needs and wishes of the mother are seen as being at best irrelevant, at worst in direct opposition, to the needs of the child.

The assumption that no birth is normal except in retrospect requires that control of the birthing process rests with those who are trained to deal with the abnormal. The medical profession, used to dealing with mechanical deviancy, have little faith in the natural processes involved in the spontaneous labour of a healthy woman. As Sally Inch has so clearly demonstrated, interventionist technology tends to be used 'just in case', but by disrupting the normal physiological pattern, further intervention may then be required to ensure a satisfactory outcome. If the process breaks down completely, surgical delivery may be necessary.

Surgical Intervention in Delivery

The case against Mrs Savage questioned her reluctance to intervene surgically during labour. Yet this question is part of an ongoing debate within obstetrics and the different viewpoints are reflected in the varying rates of Caesarean section regionally, internationally and historically. In 1981 the Office of Population Censuses and Surveys (OPCS) showed that in Yorkshire the rate was 7.9 per cent compared with 10.2 per cent in both Merseyside and the North-West Thames Health Regions. In 1971, the rate for England and Wales as a whole was 4.9 per cent, by 1983 it was more than twice as high at 10.1 per cent; in the United States the rate increased even more dramatically over the same period, from 5.5 per cent to 20.3 per cent. It seems highly unlikely that these figures can be explained simply in terms of relative risk factors. The World Health Organisation has stated that there is no justification for any geographic region to have more than 10 to 15 per

48

cent Caesarean births, and some obstetricians feel that this figure is still too high.

Although the risks involved in surgical delivery have been halved, the fact that twice as many are now performed means that the number of women at risk remains the same:

'The latest Confidential Enquiry into maternal mortality is the first in which the maternal mortality rate has not fallen. This has been in part due to increased rates of pulmonary embolism and following anaesthesia, ruptured uterus and haemorrhage. The overall embolism rate following Caesarean section has remained the same. However, more women are having Caesarean sections and there are therefore more pulmonary emboli. This suggests that in Britain, we have reached the point where medical intervention is causing deaths for women with no further salvage for babies.'[14]

Futhermore, all surgical procedures have risks and all are associated with high levels of morbidity such as haemorrhage, infection, pain and tiredness. Looking after a newborn baby, getting up at night, establishing breast-feeding, does not seem the ideal way to recover from surgery. Caesarean delivery may be life-saving at times, but at other times it is arguable whether a long convalescence is preferable to a long labour.

Pain in Labour

The active management of birth as practised by orthodox medicine requires that the doctor is active, not the patient. In medicalized labour the woman is usually placed in the lithotomy postition, ie. flat on her back with her legs in stirrups for delivery; may be given an intravenous infusion; denied food 'just in case' surgical delivery is later required and then strapped to a monitor. Although not all procedures are used in every case, choice and control are denied to the woman in the name of safety. Sally Inch suggests that the lithotomy position is the most common position for delivery in the western world. However, she argues that by restricting the woman's movement and her ability to actively respond during labour it in fact

49

increases discomfort, necessitating greater use of pain relieving drugs and may also lengthen the duration of labour. This was reiterated by virtually every speaker at the first International Conference on Home Birth held in London in October 1987. They each argued that a relaxing, supportive environment where the woman is allowed to be in a position which is comfortable for her will maximally facilitate the progress of labour and help her respond to her pain rather than to work against it.

Yet health-care professionals have tended to view the process primarily in terms of pain, resulting in a further justification of medical control. For example, to quote Gordon Bourne again:

> 'The word labour as a synonym for the process of delivery is unfortunate and it is equally unfortunate that as far back as recorded evidence is available it has been associated with pain, fear, anxiety and sometimes disaster. Modern medicine has ensured the welfare of a woman in labour as well as that of her child, even when complications arise, but in the vast majority of labours that are free from complications modern medical aids not only ensure the safety of the mother and her child, but also ensure that she will receive all the help and assistance that she requires.'[15]

Too often modern medical aids have superceded human contact and emotional support, so that the mother's demands for attention in labour are responded to with analgesia, reaffirming the definition both of the situation as primarily painful and of herself as a passive patient. In the 1940s obstetrician Grantly Dick-Read argued that pain in labour is worsened by fear and tension: if taught to expect pain, a woman will be frightened and therefore tense, so that pain may well be increased. Thus to reduce pain it is more important to alleviate fear and tension, and perhaps the most valuable analgesics are for the woman to feel relaxed, supported and in control of the situation.[16]

There is a difference between experiencing pain in labour and defining the entire situation as such. Obstetricians have tended towards the latter,

50

offering pharmacological solutions for this perceived problem. Indeed, this became one of the major reasons why women wished to opt for medical obstetric care, particularly in the 1950s, despite the potential hazards of such drugs for both mother and child, not least being the loss of control. Beverley Beech has suggested that this has resulted in the view that women should perhaps be able to sit in bed doing the crossword while giving birth at the same time.[17] Yet (unlike Bourne) Beech, Rothman and many mid-wives I have spoken to suggest that the word labour is an accurate description of birthing and, importantly, by conceptualizing the mother's experience as work, control would be moved to the mother. Mrs Savage has stated that:

'I think that men who are not physically going to give birth, but who are onlookers and bystanders have a feeling that they have to do something about the pain and about how labour is progressing. Whereas a woman who will or may go through the experience understands that it is a very important part of how a woman functions in life. There are worse things in life than pain. To go through the process of labour and to keep in control of it is a very important part of a woman's self-esteem.'[18]

Again it can be seen that doctors have the power of definition, but this is not simply a 'male conspiracy'. It is a logical consequence of treating pregnancy and childbirth within the framework of western scientific medicine, wherein obstetrics is a surgical speciality and pregnant women are treated like any other surgical patient.

There's no doubt that modern scientific medicine and its attendant technology have been beneficial and even life-saving for many women and their babies. What is at issue is the blanket application of such technologies to all women as routine, irrespective of clinical indications and to the detriment of psycho-social needs. This situation has occurred in the name of safety: the medicalization of childbirth stems from the belief that it is fundamentally a hazardous event, where women and their babies are at risk of their lives. Yet:

> 'Unsafe or risk is a concept, not a fact and is based on the premise that all births are pathological or potentially pathological. The antithetical premise that all births are normal, or potentially normal is much more valid since around 90 per cent of births, if reasonably managed, turn out to be normal.'[19]

Therefore the issue of safety is not as cut and dried as we are often led to believe and the constant referral to dangers in obstetrics by the medical profession perhaps relates more to their own quest for professional dominance and control. Given adequate support and information women are not going to risk their own lives or that of their baby and must have the right to good medical care and attention as and when necessary. But by defining childbirth as primarily a social event, in which the birthing woman is an active participant, and seeing pregnancy and childbirth as an integral part of the lives of healthy women, the context in which maternity services are delivered drastically changes. The dominance of a predominantly male medical profession in this arena, and indeed in relation to all aspects of fertility control, would be much reduced. This would be a step towards giving women greater self-determination over their lives.

REFERENCES
(1) W. Savage, *Observer*, 5-7-85
(2) R. Campbell and A. MacFarlane, 1987, p.4
(3) see A. Oakley, 1984
(4) Peel Committee Report, 1970, para.248
(5) J. Squire, 1986
(6) I. Chalmers et al, *BMJ*, 1976, pp.735-8
(7) Fryer and Ashford, *British Journal of Social and Preventive Medicine*, 1975, no.26, pp.1-9
(8) M. Tew, Wembley Conference, 1987
(9) R. Campbell and A. Macfarlane, 1987

(10) Inch, 1982, p.10 (foreword by P. Huntingford)
(11) B. Beech, 1987
(12) G. Bourne, 1984, p.25
(13) S. Kitzinger, *Sunday Times*, 15-5-85
(14) W. Savage, 1983, *Medicine in Society*, vol.9, no.4
(15) G. Bourne, 1984, p.17
(16) see B. Rothman, 1982
(17) B. Beech, 1987
(18) W. Savage, *Guardian*, 20-2-85
(19) M. Wagner, 1986, p.16

4

THE HEALTH SERVICE AND ITS USERS

An editorial in *The Lancet* (2-8-86) suggested that the root of the conflict between Mrs Savage and her colleagues was 'simply a difference in attitudes when looking after patients'. The use of the word 'simply' reveals how the importance of professional attitudes towards patients, the 'bedside manner', is underestimated by the medical profession itself. The paternalism inherent in the doctor/patient relationship rarely acknowledges patients' rights of autonomy in sickness and in health. While patients are expected to trust their doctors, this trust is not reciprocated; you may consent to treatment, but rarely are you given a choice. The move towards woman-centred obstetrics challenges this position, and the response of women to Mrs Savage's case shows that this difference in attitudes is actually of fundamental importance and reflects the gulf between what women are seeking in terms of obstetric care and what the medical profession wants to provide.

Yet the issue of choice and control is fundamentally linked to the level of involvement in the planning and implementation of care and therefore the woman-centred approach to maternity care seeks to actively involve women in this. The World Health Organisation endorses this approach and has stated that:

> 'Each woman has a fundamental right to receive proper prenatal care; that the woman has a central role in all aspects of this care, including participation in the planning, carrying out, and evaluation of the care; and that social, emotional and psychological factors are decisive in the understanding and implementation of proper prenatal care.'

WHO 1985

This may also be linked to the WHO's overall philosophy which encourages a community development approach to health promotion and sees the community as being the primary site for action and involvement in this area (eg. its 'Health For All by the Year 2000' programme). In so doing it emphasizes the role of the users, or patients, as being at the centre of events and thereby having some control over their health and their lives.

For Dr Marsden Wagner, Regional Organiser of the WHO Office of Maternal and Child Health, it is essential for women to feel in control throughout their pregnancy and he suggests that choice is a necessary prerequisite for this – indeed, the ability to exercise choice may be more important than the choice itself.[1] Yet such a view runs directly counter to prevailing medical philosophy where choice and control are relinquished to the medical expert, not merely on the basis of superior knowledge, but also, according to Gordon Bourne, because pregnancy itself causes personality changes to the extent that:

'...Even the most highly competent and efficient woman may find that her judgement is impaired...It would, of course, be wrong to suggest that pregnant women are incapable, but a word of warning about emotional instability should make them consider things more carefully and may prevent them from doing things and making decisions which they might subsequently consider unwise.'[2]

Doctors have argued that challenges to their authority have created an atmosphere where they are likely to practise 'defensive medicine' for fear of litigation, pointing to the rise in the number of Caesarean sections performed as evidence of this and the fact that 70 per cent of American obstetricians have been sued sometime in their careers. Wagner's answer to this is that if they insist on playing God they must answer to any natural disaster that may follow. There is a crucial distinction between informed consent and informed choice and the practice of defensive medicine reflects the profession's reluctance to relinquish control by giving women the necessary information which would allow them to make informed choices and take responsibility for their care.

It must be remembered that Mrs Savage was charged with incompetence by her *fellow professionals*. For allowing women to choose, she herself was accused of being irresponsible. For treating each woman as an individual with individual needs and potential risks, she was charged with placing them at unnecessary risk. By taking childbirth out of the realm of scientific objectivity and into the context of subjective experience, she was branded as too dangerous to continue practising. Yet her patients could only praise her care and consideration.

Women Demand to be Heard

Ann Oakley has shown that there has always been user dissatisfaction with the type of maternity services on offer, but suggests that the wider political climate of the 1960s and 1970s allowed it to find an organisational voice.[3] It seems that this dissatisfaction has become more vocal as the move towards 100 per cent hospital births involving more and more technological intervention has largely removed any notion of choice for child-bearing women. That this voice has been heard can be seen in the way in which some demands have been co-opted into the medical model, but a political acceptance of the need to listen to women does not necessarily imply a change of heart. As Beverley Beech has argued:

'The consumers are asking to be treated as responsible, sensible adults who require support and tender loving care through this event. The professionals have responded to this call by decorating. They have provided single rooms (a positive step), they have papered the walls, added flowered coverings to the bed, hidden the technology in a cupboard and then trumpeted that they have responded to the user's needs. Often this response has not included a change of attitude.'[4]

Even Gordon Bourne's book of advice for pregnant women acknowledges the rise of a vocal consumer group, his response being to explain to women the wonders of medical science, stating that it was:

> 'The feeling of all who look after pregnant women that it was high time they understood more about themselves, their pregnancies and their labours...The more the woman can be instructed, the more easily will her full cooperation be obtained.'[5]

Not only does he suggest that it was the doctor's idea to share information, but he also manages to subvert the issue. Women want information to enable them to play an active role; Bourne is providing 'instruction' so that women can become better (more passive?) patients.

In the same way some alternative childbirth methods have been accepted by professionals more than others, for example the Lamaze technique, a method of preparation for childbirth, has been widely endorsed in the United States, because once again it accepts the need to co-operate with your medical advisor. Many hospitals now use Birth Plans which allow women some say in planning their care. However, whereas in the woman-centred model this is used to enable women to carefully consider what they want, with the full support of their doctors or midwives; when it has been adopted by the medical model it is far less comprehensive – often women are asked simply to tick whether they wish/don't wish/are indifferent to a particular procedure. Clearly this reduces its value somewhat.

In 1982 the Maternity Services Advisory Committee proposed that each health district should set up a Maternity Services Liaison Committee (MSLC) to encourage professional and user co-operation in the planning of services. However this has not been particularly successful: one district has set up two committees, one involving users, one that is for professionals only, but only the latter has the power to institute change; in another district there was considerable user interest and participation, with the result that the professionals refused to attend. In evidence to the Tower Hamlets Health Inquiry, Tower Hamlets Health Campaign argued that:

> 'Consultation machinery is centrally important in planning and developing services in an area like Tower Hamlets. We are not at all

happy with the Maternity Services Liaison Committee, dominated as it is by professional medical viewpoints. Given its unanimous resolution that Mrs Savage should be replaced by a woman locum, it seems to have no clout in addition to its unrepresentativeness. There is immediate need for such a forum to discuss what sort of consultation is likely to be most effective for local women.'

It can be seen, therefore, that despite an active and vocal consumer movement concerned to change the way in which maternity services are delivered, what impact it has had has been unable to change the medical context of childbirth and it has continually come up against a well-organised and powerful medical establishment. Wagner has stated that even a group as prestigious as the World Health Organisation has had to battle against this powerful lobby. The suspension of Mrs Savage raises important questions of professional accountability, revealing how little weight is given to patients' opinions when a professional can only be judged by her peers. As Beverley Beech, Chair of AIMS and Jean Robinson, ex-Chair of the Patients' Association, stated in a letter to *The Guardian*, they have both dealt with many cases where patients have complained about their care or about medical attitudes which have not once resulted in suspension:

'The profession is noted for the way it closes ranks when a consumer makes a complaint of a colleague.'

Accountability in the NHS

Yet the issue also raises more fundamental questions concerning power in the health service, and the suspension of Mrs Savage may be linked to a range of problems experienced by health service users, such as cuts, closures, and inappropriate and inadequate services. It has made it clear that users have little power to influence decisions in the NHS.

Of particular concern in this affair is the role of Francis Cumberlege, Chair of the DHA. He played an instrumental part in establishing the case and in deciding to go for a full-scale enquiry, which he did without notifying either Mrs Savage herself or the full DHA. The first official notification received by members of the authority was when he announced at an ordinary meeting of the DHA in May 1985 that the enquiry would be taking place, two weeks after the suspension and four weeks before the time Mrs Savage had been given to respond to the allegations had expired. He then refused to discuss the issue in public, declaring it sub judice, which was not true, it being only an enquiry and not a trial. However, the absence of a legal practitioner on the DHA meant that no concerted effort to challenge this statement was made. Later he was to write to the Chair of the NHS Consultants Association stating:

'Mr Trevor Beedham, who heads the department so competently, and Professor Grudzinskas, who was brought back from Australia to be professor in obstetrics and gynaecology...together with Mr Hartgill and Mr Oram are now running a forward-looking department in complete harmony.'[6]

It would appear that, like the Professor, Cumberlege's main concern was the harmony of the department, not with ensuring that the health authority was providing services which are geared towards meeting local needs. As Chair of the DHA, he has some obligation to ensure this, but by taking the decision to deprive the community of the services of a consultant who had its full support, he obviously failed to meet the community's needs. In 1986 he was reappointed by the Secretary of State for a further two years as Chairman, thus blocking the way to a speedy and total resolution of the affair, rather than the damage-limitation steps which were in fact eventually taken. This suggests that political patronage may play a larger role in such appointments than the views of the community. Significantly in November 1987 when the authority was forced to make drastic cuts in services, Cumberlege stated that it was the duty of the DHA to support the Secretary of State.

The District Health Authority(DHA) and the Community

Since 1982 DHAs have been identified as the key accountable body, responsible for the planning and provision of local services to meet local needs.

Yet many health authorities have poor links with the communities they serve and too often local priorities for health have to compete for scarce resources against the more powerful professional lobbies, such as hospital consultants and the medical college. DHAs have no independent source of revenue and therefore are entirely dependent on central fiscal policy, as witnessed by the present financial crisis faced by most health districts. They are also appointed bodies and as such they are accountable to those who appoint them and not to the local community. Thus there is little incentive for them to look outwards, to the needs and priorities of their district, or to involve local people in the decision-making process. As a result neither the DHA as a public body nor individual members have the necessary visibility or legitimacy to act in the 'community interest', nor do they have a strong basis from which to challenge professional interests, given that professional knowledge and expertise are a source of legitimate power.

Therefore local health policy tends to centre around the provision of acute hospital-based services, rather than local needs. The findings of the Tower Hamlets Health Inquiry, published in May 1987, highlighted the need for more community-based services and for better communication and integration between those responsible for providing them. The health of the population in Tower Hamlets is directly affected by social and economic factors such as poverty, homelessness, and racism; as factors affecting the health of local residents these may be of greater concern than the availability of specialist hospital services. Yet the needs and wishes of local people continue to be given less priority than those of professionals, whether managers or doctors. For example, in 1988 cuts were made to mental health services in Tower Hamlets, already hopelessly inadequate after years of under-funding, as stated in a damning report from the Mental Health Act Commissioners following a visit the previous year. DHA members voted for these cuts on the recommendation of the General Manager, who had failed to inform them of the existence of this report.

This situation will not be enhanced by the proposals outlined in the Government's white paper, *Working For Patients*. Health authorities will be much smaller and more concerned with arranging contracts for services within their limited budgets. There will be no members nominated by the local authority or the wider trade union movement, as at present; instead they will be appointed for their business skills rather than their understanding and knowledge of the local community. In short, the NHS will be accountable primarily to those who pay for the service and not to those who use it.

Often these issues come to light only at times of crisis, such as when an authority is proposing to make sweeping cuts in services or when there is other popular resentment over a particular decision (and even then it is likely to be groups other than the health authority who publicise them). The suspension of Wendy Savage is such a case. A few powerful professionals were able to set in motion a train of events which was to cost the DHA more than £250,000 at a time of severe financial stringency and to place it in direct conflict with local service users and community health professionals. The role of health authority members was in effect merely to endorse a decision which had already been taken by the Chair, at the instigation of several consultants who had only ever discussed the case with Grudzinskas and Beedham.

The entire affair has shown that little attention is paid to local women's health needs by those whose role is to provide healthcare services. Given Mrs Savage's commitment to improving women's choices and to establishing maternity, gynaecological, and abortion services to meet women's needs, the treatment she has suffered at the hands of the DHA exemplifies the lack of consideration given to local concerns. This reflects the way in which professional interests are able to dominate the organisational structure of the NHS in a way that community needs and priorities are not. The extension of the medical audit under the white paper proposals will not improve this situation, unless some measure of 'user satisfaction' with the type and quality of service, is included. While other doctors are in the best position to judge a colleague's skill and competence to practise – although, as has been shown here, this is not beyond question – a lay person is best placed to assess whether the service provided meets the needs and wishes of those who use it.

At the present time, this government is constantly reiterating the value of user choice in relation to public services and using the market mechanism to increase 'patient power'. It may be true that greater emphasis on marketing services may generate more information about services and waiting lists, but making this type of information available does not mean that users of services will be empowered to make realistic choices about the quality of care on offer. At the end of the day GPs will continue to be gatekeepers to information and services, a role which many women have been critical of when used to restrict women's access to reproductive care and control over their fertility. The Government's proposals fail to acknowledge the power of professional expertise; that without adequate appropriate information and support it is impossible for a lay person to make a meaningful or informed choice. Nor, for obvious reasons, do they question the part played by the medical profession and medical ideology in upholding the prevailing values and beliefs of the wider society, for example, beliefs about the role of women in society.

The medical model of health is central to the structure of the health service: since its inception, hospital-based services geared towards the diagnosis and treatment of illness have taken priority to the detriment of primary health care. The relative failure of policy from successive governments over the last 20 years to reverse this trend, highlights the strength of professional interests and the interventionist framework. The implementation of the findings of the Griffiths report since 1983, strengthening the role of management, has had some impact on the way in which issues are prioritized, but this has been more concerned with the financing of services at a time of minimal growth than with developing a service which would aim towards meeting the needs of local people: towards a health service rather than an illness service. Perhaps this can be seen most clearly in an area as deprived as Tower Hamlets, where a high-tech, specialist teaching hospital is in many ways a mixed blessing.

User participation in health care requires greater involvement in the running of services, so that the services provided go some way towards meeting the needs of those who use them, rather than the needs of professional providers. Such a change would fundamentally challenge the view that health can only be understood by its relationship to disease and instead focus on the relationship between our health and the conditions of our lives. This is central to the issue of women's rights of choice and control in relation to reproductive care services and hence to women's autonomy within society.

REFERENCES

(1) M. Wagner, 1987, Wembley Conference
(2) G. Bourne, 1984, p.13
(3) A. Oakley, 1984, p.236
(4) B. Beech, 1984, p.3
(5) G. Bourne, 1984, p.20
(6) *The Health Supplement*, March 1986

5

THE POLITICS OF REPRODUCTION

'The obstetrical perspective on pregnancy and birth is held to be not just one way of looking at it but to be the truth, facts, science: other societies may have beliefs about pregnancy, but we believe our medicine has the facts. But obstetrical knowledge, like all knowledge, comes from somewhere; it has a social, historical and political context.'[1]

The 'Wendy Savage Affair' was not simply an isolated story about one woman's victimization by her male colleagues, but it exposed wider issues relating to the politics of health care, and in particular to reproductive care. The way in which western scientific medicine views pregnancy and childbirth reflects the way in which it views all aspects of health and illness, ie. as being amenable to intervention and treatment. By emphasizing its potential for disaster, medicine has legitimated its control of the process of birth. Women who fight for the right to choose therefore are not simply complaining about the sexist attitudes of mostly male professionals, nor of the paternalism inherent in professionalism itself, nor even of the dehumanizing aspects of institutionalized care, although these are all relevant. What they are challenging is the appropriateness of western medicine when applied to all aspects of human health and illness and, indeed, to anything pertaining to the human body, and in particular, the anomaly of the fact that physically healthy women are thus defined by a disease-oriented framework.

The Nature of Science
The medical profession's claim to authority and hence to resources and to its exclusive rights to practise is legitimated by particular beliefs about the

nature of scientific knowledge and the scientific basis of medicine. Within this view, science is the search for objective 'facts' which exist in the natural and social world independently of the observer. Therefore knowledge gained as a result of scientific endeavour is essentially value-free, autonomous and unchallengeable. This belief in the nature of scientific knowledge has profoundly influenced not only the natural sciences but the development of western thought as a whole and has been endowed with the status of 'common sense'. It stems from the philosophy of the seventeenth century, when the greatest advances in knowledge were in the field of mathematics and mechanical physics. Of particular influence was the philosopher Descartes. Descartes believed the world to be split between science and art, objectivity and subjectivity, reason and nature, and that the only path to knowledge was via the scientific side of the equation.

Medical science was developed, and has largely remained, within this framework. Its basis is the Cartesian model which identified a distinct relationship between mind and body, with the latter viewed as a machine to be interpreted and controlled. Thus medical activities rest on the concept that human health depends on a mechanistic approach based on an understanding of the structure and function of the body and the disease processes which affect it. In this way, health and illness are viewed as being predominantly biological in origin and categories of disease as having an objective existence which can be seen by the skilled expert. Therefore ill health represents a breakdown or deviancy in normal bodily function (ie. in the machine). Relatively little attention is paid to social or environmental influences on our health.

This model has tended to give rise to the belief that biological activity and medical activity are largely synonymous. This assumed corollary and the concern with 'normality' has had particular implications for women's health care, a point to which I will return later.

The March of Progress
Over the last two decades several writers have challenged the assumption

of objectivity in medical theory and practice.[2] While recognising the undoubted benefits which have resulted from this mechanistic approach, it has also led to a limited understanding of health and illness in western society, one that is compatible with the social relations of patriarchal capitalism. Medical knowledge and techniques are not merely the logical consequence of scientific progress but rather the product of particular social, economic and political forces. By focussing on physiological disturbances in the individual patient, medical science ignores the more subjective and social dimensions of health and illness, so that, 'we come to believe we have as little control over our bodies and our health as we have over other aspects of our lives.'[3]

Usually, the impression we are given is that the advancement of medical science has been a march of progress through history as a consequence of scientific innovation and technological 'breakthroughs'. This view has gained credence with the advance of technology and biochemical knowledge in the twentieth century and is reinforced by the wide publicity given to procedures such as organ transplants and *in vitro* fertilization techniques. Yet a closer look reveals that many of the improvements in the health of the nation occurred before the advancement of medical knowledge. For example, control of infectious diseases in the early part of this century was largely due to public health measures and improved standards of living, such as better housing and nutrition. Even today, there is considerable evidence as to the influence of social and economic factors on the incidence and experience of ill health.[4] The individualistic, mechanistic approach of modern medicine obscures these connections, so that the problems of ill health are seen in isolation from the social organisation of the society in which they arise and thus are deprived of any inherent political content.

Furthermore, the history of medicine has been one of increasing patriarchal control.[5] Traditionally all areas of healing and fertility control had been dealt with by women, for women, within the context of mutual aid and self help within the community; but with the emergence of medicine as a male discipline there was a gradual transition from female to male control. This 'take-over' cannot be explained as the result of scientific advances by formally trained physicians replacing the 'Old Wives' Tales' of the female lay healer. For these women had a recognised skill status within their community, using knowledge based on empirical observation, often passed down through generations, which included the use of herbal remedies and midwifery skills. Their male counterparts on the other hand relied heavily on theology and superstition, rather than empiricism, and dubious practices such as blood-letting. Yet throughout the seventeenth century, women healers were accused of witchcraft and the expert testimony of male physicians was used in evidence against them. Thus the 'art' of healing became replaced by the 'science' of medicine.

Given that it had little actual therapeutic value at this time, it would appear that patriarchal ideology was a more crucial factor in this transition than scientific knowledge. As Ann Oakley argues, the existence of women healers challenged three social hierarchies: that of church over laity; landlord over peasant; man over woman.[6] Significantly, the Cartesian concept of 'pure' science identified objectivity and reason as being masculine; women were supposed to embody subjectivity, emotionality and nature. Thus this conception of science is inherently patriarchal – and given that Descartes believed that through science rational man could control nature, it contains within it the assumption of the domination of women by men.

As science took over from religion as the main way of understanding the world in the nineteenth century, medicine itself became a powerful social force. Its claim to objectivity and factual knowledge not only added weight to its pronouncements but also contributed to a justification of social and sexual divisions within society. At this time medical science played an important role in upholding Victorian values, in particular by defining womanhood as inherently pathological. Ruled by their uterus and ovaries, the basis of all 'female complaints', middle-class women were expected neither to possess nor display sexuality, but to devote their energy to motherhood. Too much mental or physical activity would damage their reproductive capabilities so it was imperative they confined themselves to the private sphere of social life. Working class women were obviously not defined in these terms, but were themselves seen as potentially sickening to the middle classes, particularly to middle-class men through prostitution. (The high incidence of tuberculosis among prostitutes was said to be the result of their 'immoral lifestyle'.)

In the twentieth century, medicine has tended to define women as being mentally, rather than physically, inferior and psychological pronouncements concerning womanhood, motherhood and female sexuality, particularly those of Freud, have had an important influence. In 1972, a study undertaken in the USA showed how professional definitions of the normal healthy adult tend to be equated with the adult male, so that women are implicitly seen as being other than normal.

The Medicalization of Womanhood

The assumption that medical activity and biological activity are largely synonymous has meant that women's normal biological and reproductive processes are seen to be a legitimate area of medical concern. However, contemporary medicine continues to associate women's biological role, ie. their ability to bear children, with their social role, as wife and mother and by presenting this as scientific fact plays an important ideological role, by justifying gender differences and the sexual division of labour as 'natural'. For example, the following is taken from Gordon Bourne's book *Pregnancy*:

> 'If a particular species is going to survive, then the vital functions performed by that species which are important for its survival must give a certain amount of pleasure: thus we find that human beings find pleasure in eating, pleasure in drinking, pleasure for the male in providing for his female and pleasure for the woman in looking after her man...
>
> The intensely feminine female ought to be the ideal human reproductive machine and such a person will usually attract and be attracted

to the masculine type of male, so that by a process of natural selection at the biological level the ideal reproductive female is mated with the ideal reproductive male...'[7]

Bourne also suggests that from a 'purely biological' aspect, 'masculine' women and effeminate men are not conducive to the procreation of the human race. Fortunately, however, 'nature caters for this phenomenon' by introducing homosexuality:

'Associations of this nature are, of course, reproductively sterile which is the obvious biological solution to the problem insofar as it automatically eliminates such particular types from procreation.'[8]

Obviously Bourne has never knowingly cared for lesbian mothers! In this way cultural definitions of femininity, sexuality and reproduction are reflected and reinforced by the supposedly objective and autonomous knowledge base of medical science. Femininity is seen as being inextricably linked to reproductive function and its assumed corollary, motherhood. Yet this 'dominant reproductive ideology' is not only a belief about sex and gender, but about marriage as well, so that for all married women motherhood is perceived as a desired step towards fulfilling their natural destiny, but is undesirable for single women.[9] Therefore, both the presentation of medical services and the presumptions on which they are based do not simply reflect medical knowledge and expertise, as John Ehrenreich argues:

'The scientific knowledge of doctors is sometimes not knowledge at all, but rather social messages (eg. about the proper behaviour of women) wrapped up in technical language. And above all both the doctor-patient relationship and the entire structure of medical services are not mere technical relationships but social relationships which express and reinforce (often in subtle ways) the social relations of the wider society.'[10]

By the same token abortion is not considered a valid option for married women and there is evidence to suggest that it is easier to obtain for black and single working class women. Furthermore, women who would prefer to choose abortion are treated with less sympathy than infertile married couples and while abortion services are given low priority, enormous resources are put into the development of sophisticated technologies such as *in vitro* fertilization. Even so, infertility services are not easily available, and, despite being a major concern for many women, were very poor until *in vitro* techniques came along – and men could then control them. Yet just as the social status of some women may make it easier for them to have an abortion, so will the status of others make it easier for them to gain access to such techniques. In October 1987, it was reported that a woman who had been undergoing infertility treatment was denied *in vitro* fertilization because she had previously been a prostitute, a decision that was upheld in court.

Women, Medicine and Society

The intervention of medicine in this arena is inextricably linked with the wider social control of reproduction and of women's lives in capitalist society. A division of labour, between those who produce goods and services in the market and those undertaking reproductive and domestic tasks in the private sphere of social life, is integral to the production of capital, which is dependent on exploitative relationships. Patriarchal ideology is one way of defining the structure of such relationships. The interconnection between this ideology and capitalism has ensured that women's capacity to bear children has been interpreted as *the* biological difference between women and men and hence as an *explanation* of the sexual division of labour (and of male domination). Women's primary task within capitalism is to reproduce and therefore some form of control over this task becomes necessary. The emphasis on biological difference has meant that control of reproduction has fallen to those who perceive any-thing pertaining to the human body as falling within their remit, ie. to the medical profession. Therefore the issue of women's rights of choice and control in relation to reproductive care services contains within it a direct challenge both to the autonomy of medical science and to the position of women in western society. As Lesley Doyal has argued:

'...while much sexual repression is clearly related to an ideology of

70

male domination, it is also clear that if social relations are to remain basically unchanged and capital accumulation to be maximally facilitated, then some degree of social control has to be exerted over women's sexuality, over who gives birth and under what conditions.'[11]

For all women the ability to control their fertility, to decide if, when and where they wish to give birth, is a fundamental prerequisite for them to control not just their sexual and biological lives, but also their social and economic roles. At present access to the information, technology and services which would enable women to have this self-determination is controlled by the (mostly male) medical profession. Thus the presumptions on which these services are based directly affect women's rights and choice in this area and hence the amount of autonomy they have over their lives. By actively contributing to the creation and maintenance of discriminatory views of women, medical knowledge and practice contributes to the maintenance of the sexual division of labour characteristic of 'advanced' capitalist society.

The development of modern maternity care has been seen as a legitimate area of state interest, as well as medical interest. However, as with other areas of social policy, this has been less concerned with the status of women's health as such, than with the future health of the nation – the need for healthy women to produce and care for healthy children. Therefore state intervention in this arena is part of the wider social control of women's sexuality and is facilitated by a male dominated medical profession whose power is derived from its role as an agency of the state. While the state organises the social, political and economic context in which health care services are delivered, the medical profession has technical autonomy and as medical practice becomes increasingly technological, medical influence over the lives of patients is substantial.[12] Its control over the ideology and technology of reproduction, means its influence over the lives of healthy women is more substantial still.

Mrs Savage, in common with a small number of other obstetricians, believes that women should be the central focus of reproductive care and should be given the opportunity to decide for themselves if, when and where they want to give birth. She recognises the importance for women to be able to control their own fertility – and indeed, their sexuality – and the implications this has for other aspects of women's lives in western society. This is why so many women, not just Mrs Savage's patients, but women up and down the country, came to her support. Yet perhaps the greatest irony of this affair is the fact that in June 1985, two months after her suspension, Mrs Savage was awarded the Fellowship of the Royal College of Obstetricians and Gynaecologists, which, as Nicholas Timmins stated in the *Independent*, 'is an honour bestowed neither lightly nor on one with the reputation of being incompetent'.

REFERENCES
(1) B. Rothman, 1982, p.33
(2) eg. I. Illich, 1976; J. Ehrenreich, 1978; Renaud, 1978; L. Doyal and I. Pennell, 1979
(3) L. Doyal and I. Pennell, 1979
(4) eg. *The Black Report*, 1980; *The Health Divide*, 1987
(5) eg. A. Oakley, 1976; M. Wagner, 1986
(6) A. Oakley, 1976
(7) G. Bourne, 1984, p.33
(8) G. Bourne, 1984, p.34
(9) see J. Busfield, 1974; S. MacIntyre, 1977
(10) J. Ehrenreich, 1978, p.15
(11) L. Doyal, 1983, p.33
(12) see A. Oakley, 1984

CONCLUSION

'The consumer revolt has fought, and continues to fight against prevailing medical definitions of pregnancy and antenatal care for the very sound reason that what pregnant women have complained about is the capturing of the womb by medical professionals. The consumer's focus has been the reduction of the social and personal experience and the individuality of pregnant women, to the mechanical image of the womb, housed in the body of either a reluctant or compliant patient, and processed on the principle that a no-risk birth is only to be achieved by exposing all wombs and their owners to an identical all-risk monitoring process.'[1]

In all aspects of health care there is a fundamental difference in the perception of health and illness between the providers and recipients of health care services: for those who practise medicine it is considered to be an objective, scientific endeavour; for those who suffer ill health it is a subjective experience which impinges on the rest of their lives. The dominance of the mechanistic model of health that is concerned with individual physiology and the disease process has been at the expense of the experiential dimension (the subjective feelings of the patient), to the detriment of both. The close association between this model and the structure of medical services has been crucial. It has allowed the medical profession to establish control and with it occupational prestige and the rewards, both material and symbolic, which are thus accrued. Their claim to authority and to resources rests on an ideology of altruistic service based on rational, scientific, objective knowledge. Thus medical science provides not just the only viable means for mediating between people and disease but also the only legitimate definitions of what is health and what is pathology. But this approach is severely limited: it means that our understanding of health is in its relationship to disease and therefore is reduced to merely the absence of illness. Secondly, it means that illness itself is characterized by the sum of its parts. The concern with objectivity has obscured the fact that medical activity is a human activity and that the relationship between practitioner and client, a relationship which is central to medical practice, is actually subjective, as is all human interaction.

This can be seen most clearly in relation to reproductive care services, where physically healthy women are defined and treated within a disease-oriented framework. For expectant mothers, pregnancy and birth is a major life event marking their transition to parenthood, which implies the achievement of a status of social seniority and responsibility. But through-out this time she must also take on the subordinate and passive role of patient, which implies relinquishing responsibility in the face of profes-sional expertise. Maternity care is something that is done to women. It is done on the assumption that the natural process of birth is akin to illness; it is seen as a pattern of deviancy within the normal functioning of the machine. This ignores the fact that for pregnant women the physiological changes which occur are part of a normal pattern within the 'female ma-chine'. This is not to suggest that malfunction does not occasionally hap-pen, this occurs even in the non-pregnant population, but merely that risk is not a label that needs to be attached to all pregnancies.

Yet on this basis of risk doctors have developed a system of technological intervention which it applies to all, 'just in case'. This includes induction of labour, routine ultrasound, routine episiotomy, routine electronic fetal monitoring, elective (ie. planned) forceps deliveries and Caesarean section for breech presentation and low birth weight babies. Yet none of these procedures have been subjected to proper scientific evaluation to establish the extent to which they are effective, or to which they may actually con-tribute to morbidity. What they have effectively contributed to is the predominance of medical control over the birthing process and as Rothman suggests, to manage or to control a situation is to manage or control indi-viduals. Thus the medical management of birth is the management of birthing women.

The extent to which medicine has come to dominate the field of reproduction has meant that many of the gains women have made with regard to access to information and technologies have been accompanied by a corresponding increase in medical control over this area of their lives.[2] Whether this be in terms of contraception, abortion or maternity care, women's right to choose is only valid within a medical context.

Yet an ideology of motherhood is central to theories regarding the position of women in capitalist society. Whether we have children or not, our lives are primarily defined by our capacity to bear children, socially, economically and politically. Therefore our inability to independently control our fertility has repercussions on all aspects of our lives, and given that fertility control is defined as a 'medical problem', the implication is that women's lives are medically problematic. The medicalization of reproduction is thus the medicalization of womanhood. Furthermore, by viewing pregnancy as pathology and placing birth within hospitals, motherhood itself is taken out of the realm of the public sphere of social life to become a privatised, medical event.

The suspension of Wendy Savage brought some of these issues to public attention and revealed how threatened some obstetricians feel when faced with a colleague prepared to challenge the assumptions on which they base their practice. Mrs Savage herself is perhaps only important insofar as she was the central figure in this debate and for the fact that she was strong enough to be willing to fight not simply for her professional career but for the rights of women to be centrally involved in their reproductive care. That is not to denigrate the significance of her contribution, nor to dismiss what she had to lose by this conflict, but it is essential that the issue is not seen in terms of a 'lone pioneer'. As was stated in the findings of the Enquiry Report, Mrs Savage's actions were validated by an accepted school of medical thought; while they were perhaps unusual they were within the bounds of accepted medical practice.

Possibly more important was the response of women, particularly in Tower Hamlets but nationally as well, who recognised that there was more at stake than one doctor's career. For many women the idea that doctors had different views about childbirth practices was novel and enabled them to question the policy in their hospital. In Tower Hamlets, Mrs Savage has been instrumental in developing efficient, caring abortion services and community obstetric care; services which have aimed to meet women's needs, rather than expecting women to fit into the medical model. Her suspension revealed just how little a say users have over their health services, and while this particular conflict may no longer be in the lime-light, the issue of accountability is still as crucial. With health authorities being forced to make drastic cuts in services to keep within their budgets, or making preparations to opt out of the NHS, it is perhaps more important than ever that those of us who use health services make ourselves heard.

REFERENCES
(1) A. Oakley, 1984, p.249
(2) L. Doyal, 1983

REFERENCES

Allsop, J. 1985, *Health Policy and the National Health Service*, Longmans, Essex.

Beech, B. 1984, 'Perinatal Services and Prenatal Care – The User Perspective', Health Rights, London.

Beech, B. 1987, *Who's Having Your Baby*, Camden Press, London.

Bourne, G. 1984 (2nd Ed.), *Pregnancy*, Pan Books, London.

Campbell, R. and MacFarlane, A. 1987, *Where to be Born*, National Perinatal Epidemiology Unit, Oxford.

Claxton, R. (Ed), *Birth Matters*, Unwin, London.

Doyal, L. and Pennell, I. 1979, *The Political Economy of Health*, Pluto Press, London.

Doyal, L. 1983, Women, Health and the Sexual Division of Labour, in *Critical Social Policy*, Issue 7, Vol 3, No 1.

Ehrenreich, J. 1978, *The Cultural Crisis of Modern Medicine*, Monthly Review Press, New York.

Graham, H. and Mckee, L. 1979, *The First Months of Motherhood*, H.M.S.O., London.

Graham, H. and Oakley, A. 1981, Competing Ideologies of Reproduction, in Roberts (Ed), *Women Health and Reproduction*.

Haywood, S. and Alaszewski, A. 1980, *Crisis in the Health Service*, Croom Helm, London.

Health Education Council, 1987, *The Health Divide*, H.M.S.O., London.

Hearn, J. 1985, Patriarchy, Professionalisation and the Semi-professions, in Ungerson (Ed.), *Women and Social Policy*, Macmillan, London.

Hunter, D. 1984, Managing Health Care, in *Journal of Social Policy and Administration*, Vol.18, No.1.

Illich, I. 1976, *Medical Nemesis*, Bantam Books, New York.

Inch, S. 1982, *Birthrights*, Hutchinson, London.

Jarman, B. et al., 1987, *The Tower Hamlets Health Inquiry*, Tower Hamlets Community Health Council, London.

Joffe, M. 1985, 'Foreseeable Developments in Antenatal Care in Tower Hamlets', unpublished paper to Tower Hamlets DHA.

MacIntyre, S. 1977, *Single and Pregnant*, Croom Helm, London.

Maternity Services Advisory Committee, 1983, *Maternity Care in Action*, H.M.S.O., London.

McKeown, T. 1978, *The Role of Medicine*, Blackwell, Oxford.

Mitchell, J. 1984, *What's to be Done about Health and Illness*, Penguin, Harmondsworth.

Oakley, A. 1976, Wise Women and Medicine Men, in Mitchell, J. and Oakley, A. (Eds.), *The Rights and Wrongs of Women*, Penguin, Harmondsworth.

Oakley, A. 1981, *From Here to Maternity*, Penguin, Harmondsworth.

Oakley, A. 1984, *The Captured Womb*, Blackwell, Oxford.

Rakusen, J. and Davidson, N. 1982, *Out of Our Hands*, Pan Books, London.

Regan, D. and Stewart, J. 1982, An Essay in the Government of Health, in *Journal of Social Policy and Administration*, Vol.16, No.1.

Roberts, H. (Ed.) 1981, *Women, Health and Reproduction*, Routledge and Kegan Paul, London.

Rothman, B.K. 1982, *In Labour, Women and Power in the Birthplace*, Junction Books, London.

Savage, W. 1986, *A Savage Enquiry*, Virago, London.

Squire, J. 1986, The Issues of Choice and Safety, in Claxton (Ed.), *Birth Matters*.

Stewart, N. 1986, Obstetric Drugs and Technology, in Claxton (Ed.), *Birth Matters*.

Townsend, P. and Davidson, N. 1982, *Inequalities in Health: The Black Report*, Penguin, Harmondsworth.

Wagner, M. 1986, The Medicalisation of Birth, in Claxton (Ed.), *Birth Matters*.

World Health Organisation, 1985, 'Summary Report on the Joint Inter-regional Conference on Appropriate Technology for Birth'.

First International Conference on Home Birth, Wembley Conference Centre, 23/24-10-87. Speakers including: Rona Campbell, Ann Oakley, Michel Odent, Wendy Savage, Marjorie Tew, Marsden Wagner, Luke Zander

health rights

Health Rights is a London-based voluntary organisation working on a broad range of public health and NHS issues. It carries out research; publishes reports and booklets; organises conferences and campaigns; works with a wide range of organisations and agencies. Health Rights provides advice and information to individuals, voluntary and community groups, health professionals, statutory bodies, and the media. It receives core funding from the London Boroughs Grants Scheme.

Aims and objectives

Health Rights believes that everyone, irrespective of sex, race, class, sexual orientation, age, physical and mental ability, has the right to the health care and treatment they need and the right to health. Health Rights' aim is to promote these basic 'health rights'. It believes they can only be achieved through substantial improvements in the nature, quality, and organisation of health services; and through more comprehensive and effective public health measures which prevent avoidable illness and disability.

To achieve these aims Health Rights works on a number of key NHS issues to secure specific improvements in health services, and more generally, promotes the need for greater commitment to preventive and public health measures.

Further information is available from: Health Rights, Unit 110, Bon Marche Building, 444 Brixton Road, London SW9 8EJ.